Innovation, Regional Development and the Life Sciences

The life sciences is an industrial sector that covers the development of biological products and the use of biological processes in the production of goods, services and energy. This sector is frequently presented as a major opportunity for policy-makers to upgrade and renew regional economies, leading to social and economic development through support for high-tech innovation.

Innovation, Regional Development and the Life Sciences analyzes where innovation happens in the life sciences, why it happens in those places, and what this means for regional development policies and strategies. Focusing on the UK and Europe, its arguments are relevant to a variety of countries and regions pursuing high-tech innovation and development policies. The book's theoretical approach incorporates diverse geographies (e.g. global, national and regional) and political-economic forces (e.g. discourses, governance and finance) in order to understand where innovation happens in the life sciences, where and how value circulates in the life sciences, and who captures the value produced in life sciences innovation.

This book will be of interest to researchers, students and policy-makers dealing with regional/local economic development.

Kean Birch is an Associate Professor in the Department of Social Science at York University, Canada.

Regions and Cities

In today's globalised, knowledge-driven and networked world, regions and cities have assumed heightened significance as the interconnected nodes of economic, social and cultural production, and as sites of new modes of economic and territorial governance and policy experimentation. This book series brings together incisive and critically engaged international and interdisciplinary research on this resurgence of regions and cities, and should be of interest to geographers, economists, sociologists, political scientists and cultural scholars, as well as to policy-makers involved in regional and urban development.

For more information on the Regional Studies Association visit www.regionalstudies.org

There is a **30% discount** available to RSA members on books in the ***Regions and Cities*** series, and other subject related Taylor and Francis books and e-books including Routledge titles. To order just e-mail Cara.Trevor@tandf.co.uk, or phone on +44 (0) 20 7017 6924 and declare your RSA membership. You can also visit www.routledge.com and use the discount code: **RSA0901**

106 Governing Smart Specialisation
Edited by Dimitrios Kyriakou, Manuel Palazuelos, Inmaculada Periañez-Forte, and Alessandro Rainoldi

105 Innovation, Regional Development and the Life Sciences
Beyond clusters
Kean Birch

104 Unfolding Cluster Evolution
Edited by Fiorenza Belussi and Jose Luis Hervás-Olivier

103 Place-based Economic Development and the New EU Cohesion Policy
Edited by Philip McCann and Attila Varga

102 Transformation of Resource Towns and Peripheries
Political economy perspectives
Edited by Greg Halseth

101 Approaches to Economic Geography
Towards a geographical political economy
Ray Hudson

100 Secondary Cities and Development
Edited by Lochner Marais, Etienne Nel and Ronnie Donaldson

99 Technology and the City
Systems, applications and implications
Tan Yigitcanlar

98 Smaller Cities in a World of Competitiveness
Seter Karl Kresl and Daniele Ietri

97 Code and the City
Edited by Rob Kitchin and Sung-Yueh Perng

96 The UK Regional–National Economic Problem
Geography, globalisation and governance
Philip McCann

95 Skills and Cities
Edited by Sako Musterd, Marco Bontje and Jan Rouwendal

94 Higher Education and the Creative Economy
Beyond the campus
Edited by Roberta Comunian and Abigail Gilmore

93 Making Cultural Cities in Asia
Mobility, assemblage, and the politics of aspirational urbanism
Edited by Jun Wang, Tim Oakes and Yang Yang

92 Leadership and the City
Power, strategy and networks in the making of knowledge cities
Markku Sotarauta

91 Evolutionary Economic Geography
Theoretical and empirical progress
Edited by Dieter Kogler

90 Cities in Crisis
Socio-spatial impacts of the economic crisis in Southern European cities
Edited by Jörg Knieling and Frank Othengrafen

89 Socio-Economic Segregation in European Capital Cities
East meets West
Edited by Tiit Tammaru, Szymon Marcińczak, Maarten van Ham and Sako Musterd

88 People, Places and Policy
Knowing contemporary Wales through new localities
Edited by Martin Jones, Scott Orford and Victoria Macfarlane

87 The London Olympics and Urban Development
The mega-event city
Edited by Gavin Poynter, Valerie Viehoff and Yang Li

86 Making 21st Century Knowledge Complexes
Technopoles of the world revisited
Edited by Julie Tian Miao, Paul Benneworth and Nicholas A. Phelps

85 Soft Spaces in Europe
Re-negotiating governance, boundaries and borders
Edited by Philip Allmendinger, Graham Haughton, Jörg Knieling and Frank Othengrafen

84 Regional Worlds: Advancing the Geography of Regions
Edited by Martin Jones and Anssi Paasi

83 Place-making and Urban Development
New challenges for contemporary planning and design
Pier Carlo Palermo and DavidePonzini

82 Knowledge, Networks and Policy
Regional studies in postwar Britain and beyond
James Hopkins

81 Dynamics of Economic Spaces in the Global Knowledge-based Economy
Theory and East Asian cases
Sam Ock Park

80 Urban Competitiveness
Theory and practice
Daniele Letri and Peter Kresl

79 Smart Specialisation
Opportunities and challenges for regional innovation policy
Dominique Foray

78 The Age of Intelligent Cities
Smart environments and innovation-for-all strategies
Nicos Komninos

77 Space and Place in Central and Eastern Europe
Historical trends and perspectives
Gyula Horváth

76 Territorial Cohesion in Rural Europe
The relational turn in rural development
Edited by Andrew Copus and Philomena de Lima

75 The Global Competitiveness of Regions
Robert Huggins, Hiro Izushi, Daniel Prokop and Piers Thompson

74 The Social Dynamics of Innovation Networks
Edited by Roel Rutten, Paul Benneworth, Dessy Irawati and Frans Boekema

73 The European Territory
From historical roots to global challenges
Jacques Robert

72 Urban Innovation Systems
What makes them tick?
Willem van Winden, Erik Braun, Alexander Otgaar and Jan-Jelle Witte

71 Shrinking Cities
A global perspective
Edited by Harry W. Richardson and Chang Woon Nam

70 Cities, State and Globalization
City-regional governance
Tassilo Herrschel

69 The Creative Class Goes Global
Edited by Charlotta Mellander, Richard Florida, Bjørn Asheim and Meric Gertler

68 Entrepreneurial Knowledge, Technology and the Transformation of Regions
Edited by Charlie Karlsson, Börje Johansson and Roger Stough

67 The Economic Geography of the IT Industry in the Asia Pacific Region
Edited by Philip Cooke, Glen Searle and Kevin O'Connor

66 Working Regions
Reconnecting innovation and production in the knowledge economy
Jennifer Clark

65 Europe's Changing Geography
The impact of inter-regional networks
Edited by Nicola Bellini and Ulrich Hilpert

64 The Value of Arts and Culture for Regional Development
A Scandinavian perspective
Edited by Lisbeth Lindeborg and Lars Lindkvist

63 The University and the City
John Goddard and Paul Vallance

62 Re-framing Regional Development
Evolution, innovation and transition
Edited by Philip Cooke

61 Networking Regionalised Innovative Labour Markets
Edited by Ulrich Hilpert and Helen Lawton Smith

60 Leadership and Change in Sustainable Regional Development
Edited by Markku Sotarauta, Ina Horlings and Joyce Liddle

59 Regional Development Agencies: The Next Generation?
Networking, knowledge and regional policies
Edited by Nicola Bellini, Mike Danson and Henrik Halkier

58 Community-based Entrepreneurship and Rural Development
Creating favourable conditions for small businesses in Central Europe
Matthias Fink, Stephan Loidl and Richard Lang

57 Creative Industries and Innovation in Europe
Concepts, measures and comparative case studies
Edited by Luciana Lazzeretti

56 Innovation Governance in an Open Economy
Shaping regional nodes in a globalized world
Edited by Annika Rickne, Staffan Laestadius and Henry Etzkowitz

55 Complex Adaptive Innovation Systems
Relatedness and transversality in the evolving region
Philip Cooke

54 Creating Knowledge Locations in Cities
Innovation and integration challenges
Willem van Winden, Luis de Carvalho, Erwin van Tujil, Jeroen van Haaren and Leo van den Berg

53 Regional Development in Northern Europe
Peripherality, marginality and border issues
Edited by Mike Danson and Peter De Souza

52 Promoting Silicon Valleys in Latin America
Luciano Ciravegna

51 Industrial Policy Beyond the Crisis
Regional, national and international perspectives
Edited by David Bailey, Helena Lenihan and Josep-Maria Arauzo-Carod

50 Just Growth
Inclusion and prosperity in America's metropolitan regions
Chris Benner and Manuel Pastor

49 **Cultural Political Economy of Small Cities**
Edited by Anne Lorentzen and Bas van Heur

48 **The Recession and Beyond**
Local and regional responses to the downturn
Edited by David Bailey and Caroline Chapain

47 **Beyond Territory**
Edited by Harald Bathelt, Maryann Feldman and Dieter F. Kogler

46 **Leadership and Place**
Edited by Chris Collinge, John Gibney and Chris Mabey

45 **Migration in the 21st Century**
Rights, outcomes, and policy
Kim Korinek and Thomas Maloney

44 **The Futures of the City Region**
Edited by Michael Neuman and Angela Hull

43 **The Impacts of Automotive Plant Closures**
A tale of two cities
Edited by Andrew Beer and Holli Evans

42 **Manufacturing in the New Urban Economy**
Willem van Winden, Leo van den Berg, Luis de Carvalho and Erwin van Tuijl

41 **Globalizing Regional Development in East Asia**
Production networks, clusters, and entrepreneurship
Edited by Henry Wai-chung Yeung

40 **China and Europe**
The implications of the rise of China as a global economic power for Europe
Edited by Klaus Kunzmann, Willy A. Schmid and Martina Koll-Schretzenmayr

39 **Business Networks in Clusters and Industrial Districts**
The governance of the global value chain
Edited by Fiorenza Belussi and Alessia Sammarra

38 **Whither Regional Studies?**
Edited by Andy Pike

37 **Intelligent Cities and Globalisation of Innovation Networks**
Nicos Komninos

36 **Devolution, Regionalism and Regional Development**
The UK experience
Edited by Jonathan Bradbury

35 **Creative Regions**
Technology, culture and knowledge entrepreneurship
Edited by Philip Cooke and Dafna Schwartz

34 **European Cohesion Policy**
Willem Molle

33 **Geographies of the New Economy**
Critical reflections
Edited by Peter W. Daniels, Andrew Leyshon, Michael J. Bradshaw and Jonathan Beaverstock

32 **The Rise of the English Regions?**
Edited by Irene Hardill, Paul Benneworth, Mark Baker and Leslie Budd

31 **Regional Development in the Knowledge Economy**
Edited by Philip Cooke and Andrea Piccaluga

30 **Regional Competitiveness**
Edited by Ron Martin, Michael Kitson and Peter Tyler

29 **Clusters and Regional Development**
Critical reflections and explorations
Edited by Bjørn Asheim, Philip Cooke and Ron Martin

28 **Regions, Spatial Strategies and Sustainable Development**
David Counsell and Graham Haughton

27 **Sustainable Cities**
Graham Haughton and Colin Hunter

26 **Geographies of Labour Market Inequality**
Edited by Ron Martin and Philip S. Morrison

25 **Regional Innovation Strategies**
The challenge for less-favoured regions
Edited by Kevin Morgan and Claire Nauwelaers

24 **Out of the Ashes?**
The social impact of industrial contraction and regeneration on Britain's mining communities
Chas Critcher, Bella Dicks, David Parry and David Waddington

23 **Restructuring Industry and Territory**
The experience of Europe's regions
Edited by Anna Giunta, Arnoud Lagendijk and Andy Pike

22 **Foreign Direct Investment and the Global Economy**
Corporate and institutional dynamics of global-localisation
Edited by Jeremy Alden and Nicholas F. Phelps

21 **Community Economic Development**
Edited by Graham Haughton

20 **Regional Development Agencies in Europe**
Edited by Charlotte Damborg, Mike Danson and Henrik Halkier

19 **Social Exclusion in European Cities**
Processes, experiences and responses
Edited by Judith Allen, Goran Cars and Ali Madanipour

18 **Metropolitan Planning in Britain**
A comparative study
Edited by Peter Roberts, Kevin Thomas and Gwyndaf Williams

17 **Unemployment and Social Exclusion**
Landscapes of labour inequality and social exclusion
Edited by Sally Hardy, Paul Lawless and Ron Martin

16 **Multinationals and European Integration**
Trade, investment and regional development
Edited by Nicholas A. Phelps

15 **The Coherence of EU Regional Policy**
Contrasting perspectives on the structural funds
Edited by John Bachtler and Ivan Turok

14 **New Institutional Spaces**
TECs and the remaking of economic governance
Martin Jones, Foreword by Jamie Peck

13 **Regional Policy in Europe**
S. S. Artobolevskiy

12 **Innovation Networks and Learning Regions?**
James Simmie

11 **British Regionalism and Devolution**
The challenges of state reform and European integration
Edited by Jonathan Bradbury and John Mawson

10 **Regional Development Strategies**
A European perspective
Edited by Jeremy Alden and Philip Boland

9 **Union Retreat and the Regions**
The shrinking landscape of organised labour
Ron Martin, Peter Sunley and Jane Wills

8 **The Regional Dimension of Transformation in Central Europe**
Grzegorz Gorzelak

7 **The Determinants of Small Firm Growth**
An inter-regional study in the United Kingdom 1986–90
Richard Barkham, Graham Gudgin, Mark Hart and Eric Hanvey

6 **The Regional Imperative**
Regional planning and governance in Britain, Europe and the United States
Urlan A. Wannop

5 **An Enlarged Europe**
Regions in competition?
Edited by Louis Albrechts, Sally Hardy, Mark Hart and Anastasios Katos

4 **Spatial Policy in a Divided Nation**
Edited by Richard T. Harrison and Mark Hart

3 **Regional Development in the 1990s**
The British Isles in transition
Edited by Ron Martin and Peter Townroe

2 **Retreat from the Regions**
Corporate change and the closure of factories
Stephen Fothergill and Nigel Guy

1 **Beyond Green Belts**
Managing urban growth in the 21st century
Edited by John Herington

Innovation, Regional Development and the Life Sciences

Beyond clusters

Kean Birch

LONDON AND NEW YORK

First published 2017 by Routledge

2 Park Square, Milton Park, Abingdon, Oxfordshire OX14 4RN
52 Vanderbilt Avenue, New York, NY 10017

Routledge is an imprint of the Taylor & Francis Group, an informa business

First issued in paperback 2019

British Library Cataloguing in Publication Data
A catalogue record for this book is available from the British Library

Library of Congress Cataloging in Publication Data
A catalog record for this book has been requested

ISBN: 978-1-138-80761-7 (hbk)
ISBN: 978-0-367-87089-8 (pbk)

Typeset in Times New Roman
by Book Now Ltd, London

For Sheila and Maple

Contents

	List of figures	xiv
	List of tables	xv
	Acknowledgements	xvi
	List of abbreviations	xviii
1	Introduction: knowledge economies everywhere!	1
2	Innovation, clusters and knowledge-based commodity chains	15
3	Innovation geographies in the UK life sciences	39
4	Innovation governance in the Scottish life sciences	60
5	Innovation imaginaries in the European life sciences	86
6	Innovation financing in the global life sciences	106
7	Conclusion: innovation and regional development for whom?	126
	Index	142

Figures

3.1	Four main concentrations of the UK life sciences sector	47
3.2	The 'clustering' of codified knowledge in the UK life sciences	52
3.3	The 'clustering' of tacit knowledge in the UK life sciences	53
3.4	The 'clustering' of commercial knowledge in the UK life sciences	53
4.1	The Scottish life sciences	68
4.2	Commodity chain relationships in the Scottish life sciences	69
4.3	Geographies of commodity chain relationships in the Scottish life sciences	70
4.4	Commodity chain relationships to public sector organizations in the Scottish life sciences	71
6.1	Global biotech industry: market capitalization, revenues and profit	111
6.2	Global 'public' biotech industry by size of firms: revenues and profit/loss	112
6.3	Global biotech industry: private financial investment	114
6.4	Global IPOs	115
6.5	Change in UK life sciences firms on public markets (2007 vs. 2011)	116

Tables

2.1 Analytical benefits of GCC approach 27
2.2 Alliance-driven governance (ADG) model 30
3.1 *Knowledge–space* characteristics of the main UK life sciences concentrations 48
4.1 The alliance-driven governance in knowledge-based commodity chains 64
5.1 The European Commission's KBBE agenda 94

Acknowledgements

Considering that this book is derived from a number of research projects over more than a decade, I have a few people to thank for getting me here.

First, I would like to thank my PhD supervisors, James Simmie and Chris Hawes, at Oxford Brookes University, as well as my PhD examiners Phil Cooke and Judy Slinn.

Second, I would like to thank Andy Cumbers, Danny MacKinnon, Vlad Mykhnenko, Andy Pike, Katherine Trebeck and Geoff Whittam who I worked with while at the University of Glasgow, as well as my other colleagues in the Centre for Public Policy for Regions. Specific thanks are due to Paul Jenkins and Catherine McManus for their research assistance during this period, and to Mike Shand for producing such excellent graphics.

Third, I would like to thank my other research collaborators, most especially Les Levidow who has contributed enormously to my intellectual development. Others include Theo Papaioannou, Stefano Ponte, Matti Siemiatycki and David Tyfield.

Fourth, I would like to thank the various anonymous referees, conference or seminar participants, readers, editors, and so on who have all provided advice along the way.

A final thanks to friends and family over the years, especially to Sheila and Maple who make life so much more fun than writing a book on regional development!

I would like to acknowledge the following research funders: Department of Planning at Oxford Brookes University; Scottish Funding Council; Economic and Social Research Council (UK); Social Sciences and Humanities Research Council of Canada; and Faculty of Liberal Arts and Professional Studies at York University, Canada.

Various chapters in this book are derived from previously published work, my thanks to the publishers for permission to reproduce it here:

- Parts of Chapter 2 come from: Birch, K. (2007) Knowledge, space and biotechnology, *Geography Compass* 1(5): 1097–1117; Birch, K. (2008) Alliance-driven governance: Applying a global commodity chains approach to the UK biotechnology industry, *Economic Geography* 84(1): 83–103; and Birch, K. (2011) 'Weakness' as 'strength' in the Scottish life sciences: Institutional embedding of knowledge-based commodity chains in a less-favoured region, *Growth and Change* 42(1): 71–96.

- Parts of Chapter 3 come from: Birch, K. (2009) The knowledge–space dynamic in the UK bioeconomy, *Area* 41(3): 273–284.
- Most of Chapter 4 comes from: Birch, K. and Cumbers, A. (2010) Knowledge, space and economic governance: The implications of knowledge-based commodity chains for less-favoured regions, *Environment and Planning A* 42(11): 2581–2601.
- Most of Chapter 5 comes from: Birch, K., Levidow, L. and Papaioannou, T. (2014) Self-fulfilling prophecies of the European knowledge-based bio-economy: The discursive shaping of institutional and policy frameworks in the bio-pharmaceuticals sector, *Journal of the Knowledge Economy* 5(1): 1–18.

Abbreviations

ABRC	Advisory Board for Research Councils
ACARD	Advisory Council for Applied Research and Development
ADG	Alliance-driven governance
AIM	Alternative Investment Market
BBSRC	Biotechnology and Biological Sciences Research Council (UK)
BIA	BioIndustry Association (UK)
BIGT	Bioscience Innovation and Growth Team
BIO	Bio Industry Organisation (USA)
BTG	British Technology Group
CLSI	Clinical Laboratory Standards Institute
CPPR	Centre for Public Policy for Regions
CRO	Clinical research organization
DBF	Dedicated biotechnology firm
DETR	Department of Environment, Transport and the Regions (UK)
DNA	Deoxyribose Nucleic Acid
DTI	Department of Trade and Industry (UK)
EC	European Commission
EFPIA	European Federation of Pharmaceutical Industries and Associations
EMEA	European Medicines Agency
EMIS	Enterprise Management Incentive Scheme
EPO	European Patent Office
ETP	European Technology Platform
EU	European Union
FDA	Food and Drug Administration (USA)
GATT	General Agreement on Trade and Tariffs
GCC	Global commodity chain
GDP	Gross domestic product
GFC	Global financial crisis
GNP	Gross national product
GPN	Global production network
GVC	Global value chain
HEI	Higher education institution
HGP	Human Genome Project

ICI	Imperial Chemical Industries
IMI	Innovative Medicines Initiative
IP	Intellectual property
IPO	Initial public offering
IPR	Intellectual property rights
ISO	International Standards Organisation
JTI	Joint Technology Initiative
KBBE	Knowledge-based bio-economy
KBCC	Knowledge-based commodity chains
KBE	Knowledge-based economy
LFR	Less-favoured region
LSA	Life Science Alliance
LSE	London Stock Exchange
MLE	Medium and large-sized enterprise
MNC	Multi-national corporation
MRC	Medical Research Council (UK)
MSE	Micro and small-sized enterprise
NEB	National Enterprise Board
NEG	New economic geography
NERC	Natural and Environmental Research Council (UK)
NGO	Non-governmental organization
NHS	National Health Service (UK)
NIH	National Institutes of Health (USA)
NIS	New industrial spaces
NRDC	National Research Development Corporation
NUTS	Nomenclature of territorial units for statistics (EU)
ODPM	Office of the Deputy Prime Minister (UK)
OECD	Organisation for Economic Co-operation and Development
PPG	Planning Policy Guidance (UK)
PRO	Public research organisation
QA	Quality audit
R&D	Research and development
RAE	Research Assessment Exercise (UK)
RC	Research Council
RDA	Regional Development Agency (UK)
rDNA	Recombinant DNA
RIS	Regional innovation systems
RP	*Research Policy*
RVC	Regional venture capital
SERC	Science and Engineering Research Council (UK)
SME	Small and medium-sized enterprise
SPRU	Science Policy Research Unit
SRA	Strategic Research Agenda
STS	Science and technology studies
TIM	Territorial innovation model

TRIPS	Trade-related Aspects of Intellectual Property Rights
USPTO	United States Patent and Trademark Office
VC	Venture capital
VOC	Varieties of capitalism
WTO	World Trade Organization

1 Introduction

Knowledge economies everywhere!

Introduction

Countries in the Global North have experienced significant upheavals in their industrial structure over the last few decades; most other countries have experienced similar dramatic changes. Since the mid-20th century various national governments, especially in the Global North, have supported policies and treaties to reduce tariffs and controls on international trade and investment. As a result, countries have had to adapt to new political-economic imperatives, especially in terms of global competition. Various academic geographers like Storper (1997), Scott (2000a, 2000b) and Dicken (2003) have argued that this globalization of the world's political economy has meant that local and regional political economies are increasingly important sites for understanding national competitive advantage, itself a concept developed by the management theorist Michael Porter (1990).

In this context, declining manufacturing production, output and employment in the Global North – see Williams (1992) for example – has necessitated the re-invigoration of local and regional political economies through the development of geographical capacities relating to service and knowledge-based sectors in which innovation plays a central role. Rather than relying on manufacturing as the base of their national economies, which characterized post-Second World War Fordism (Jessop 2002), the argument goes that these countries have had to find ways to capture value from the so-called 'knowledge-based economy' or face continual decline (e.g. OECD 1996; European Council 2000). A number of commentators have lauded the emergence of this 'new economy' (Reich 1991), 'knowledge economy' (Leadbeater 1999), or 'creative economy' (Florida 2002), arguing that it is opening up new forms of organization, new forms of production and consumption, new forms of work, and so on.

The changing industrial structure of these countries has been defined in numerous ways, as just alluded to above. Stretching back to the 1970s, scholars like Daniel Bell (1973) argued that we are entering a 'post-industrial society' characterized by increasing knowledge intensity and turn towards service and professional work. More recent scholars have argued that the 1980s witnessed the transition from Fordism to Post-Fordism in which mass production and consumption were replaced with 'flexible specialization' (Piore and Sabel 1984), 'network society' (Castells 1996), or 'neo-Schumpeterianism' (Jessop 2002). Considering that scholarly debates in this area are long-standing and wide-ranging, it would be difficult to do justice to them here, but others have noted the varied and diverse ways in which this restructuring has been conceptualized. In particular, Benoit Godin (2006a) provides a detailed discussion of the many different 'buzzwords' used over the years from Bell's 'post-industrial society' through Post-Fordism to 'information society'. Similarly, Martin Sokol (2004) critically engages with a range of these concepts in his work, outlining the ambiguities and contradictions inherent in many of them.

With this in mind, it is important to note that while long-term trends show that manufacturing has declined precipitously in the Global North since the 1960s, the fact that this has been highly uneven geographically means that it might not be as simple as the dominant narrative suggests. For example, the UK started to de-industrialize in 1960, well before global competitive pressures were supposedly exerting their influence; countries like Spain and Portugal had relatively stable manufacturing sectors well into the 1990s; and even today Germany is still the world's dominant manufacturing exporter despite competition from countries like China with significantly lower labour costs (Williams 1992; Hudson 1999; Dicken 2003). It is not that manufacturing in the Global North has hollowed out and been replaced by bright, shiny new high-tech sectors as the result of global market imperatives, but rather that different countries have faced different pressures and adapted in different ways to these pressures. Moreover, this restructuring has been explicitly localized and regionalized in that it has differed geographically both across and within countries. Some regions have witnessed rising manufacturing employment, others declining manufacturing but rising services employment, and others still have simply experienced ongoing and debilitating decline overall (Birch and Mykhnenko 2009).

Economic change is ongoing, it does not end. As a consequence it is always an important object of inquiry. This book is an attempt to understand these changes by examining the emergence of one particular high-technology sector (i.e. the 'life sciences') in one particular country (i.e. the UK). It is concerned with the growing emphasis on innovation and high-tech sectors as important contributors to regional development and renewal; the emphasis on the importance of the 'local' in academic and policy debates; and what this means for understanding the relationship between geography and innovation in the so-called *knowledge-based economy* on which we all supposedly depend for our future standard of living.

What is the knowledge economy?

What is the knowledge economy? What does it mean for society? A good starting point to answer these questions are the ongoing academic and policy debates on the role of knowledge and innovation to societies in the Global North. Most countries now have specific research and innovation policies that outline the particular government's perspective on what knowledge and innovation represent, and most take a rather simplistic approach to these concepts (e.g. Government of Canada 2014; HM Treasury 2014). When I call them 'simplistic' I mean that they adopt a narrow approach to understanding the political economy of research and innovation. For example, the UK government's 2014 science and innovation policy, *Our Plan for Growth*, is premised on a particular political-economic understanding of research and innovation:

> If we are to become a flourishing knowledge economy, we have to build on our long-standing scientific advantages and innovate. But innovation requires investment. Countries around the world recognise that science and innovation is the right path for sustainable growth ... Our ability to develop and commercialise new ideas, products and services is critical to our economic future and to providing jobs. Investment in our knowledge base is a crucial challenge for both government and business.
>
> (HM Treasury 2014: 8–9)

In this policy narrative, research produces knowledge which is then turned into innovation – defined as new products and services – in a linear pathway. As many scholars have noted, however, this linear perspective of research and innovation is deeply problematic on a number of fronts (e.g. Gibbons *et al.* 1994; von Hippel 2005; Godin 2006b; Felt *et al.* 2007). Here I mention two problems with the linear view, but there are others.

First, knowledge has become a big deal in academic and policy debates about industrial restructuring because it is often thought of as a *social* or *public good* with few limits to its use; it is frequently treated as a *commons* on which we can all draw equally and seamlessly. As critical scholarship by the likes of Philip Mirowski (2011) and David Tyfield (2012a, 2012b) has shown, this perception pervades the work of early economists of science (e.g. Robert Solow, Kenneth Arrow) and has been particularly influential ever since (also see Godin 2006b). Subsequent work on the economics of science has reinforced these assumptions, largely for political reasons relating to the legitimation of rising science spending post-Second World War. More recent literature outside of mainstream economics has problematized the idea that knowledge is freely available, easily acquired, and easily integrated into organizational processes and routines (e.g. Cohen and Levinthal 1990; Howells 2002; Nightingale 2003). In part, this later literature has focused conceptually on the problem of *tacit knowledge*, a distinction originally made by Michael Polanyi (1958) to distinguish between formal, explicit and codified forms of knowledge (e.g. patent) and embodied, experiential and non-verbal forms of knowledge (e.g. the ability to balance). In this sense, the knowledge economy is characterized by the interaction between different forms of knowledge, as well as different attempts to capture that knowledge in useful forms.

Second, knowledge is not as easy to define as policy-makers would like, nor is the notion of a knowledge economy as simple as it sounds. A number of scholars conceptualize knowledge and the knowledge economy in ways that contradict the linear view mentioned above. Several examples spring to mind here: first,

Lundvall (1992) argues that knowledge is simply a resource, while learning is the process we should focus on when considering the emergence of a 'knowledge economy'; second, Ibert (2007) argues that it is better to theorize knowledge in terms of 'knowing' since this focuses on social practice rather than more abstract entities (e.g. patent); and third, Fuller (2002), in his work on social epistemology, argues that knowledge is a social process, rather than something that originates inside someone's head. It is social in the sense that it necessitates social institutions (e.g. language) to undertake in the first place; social relations to be shared (e.g. dissemination); social practices to be produced (e.g. collaboration); and so on. As all these examples suggest, knowledge can be better thought of as a range of entangled social practices, processes and performances; consequently, the knowledge economy is better thought of as similarly messy – that is, collaborative, competitive, unidirectional, bidirectional, etc. all at once.

Why innovation?

We can ask a similar set of questions when we think about innovation. What is innovation? What does it mean for society? Whatever conception of knowledge or the knowledge economy we ascribe to, most people view innovation as an unalloyed social good. By this I mean that most people now see innovation as *the* source of economic, social *and* ecological development. This assumption is most obvious in the European Union's (EU) new research and innovation strategy, *Horizon 2020*, which replaces the previous *Lisbon Agenda* (2000–2010), and runs from 2010 until 2020 (Birch and Mykhnenko 2014). The new EU strategy is sub-titled 'a strategy for smart, sustainable and inclusive growth' and includes flagship programmes like 'Innovation Union' designed:

> ... to improve framework conditions and access to finance for research and innovation so as to ensure that innovative ideas can be turned into products and services that create growth and jobs.
>
> (CEC 2010: 3)

In this policy narrative, innovation is presented as the solution to a range of societal problems, which ends up naturalizing market-based approaches to social issues as diverse as youth unemployment, social exclusion and climate change.

The reason that innovation naturalizes market conceptions of the world is because innovation is frequently conceptualized in *Schumpeterian* terms as the commercial introduction of new *technological* products and services (Freeman 1974; Simmie 2001). This definition is prevalent in the field of innovation studies, especially research following the tradition established by people like Christopher Freeman (1974) who established the Science Policy Research Unit (SPRU) at Sussex University. However, Godin (2015) argues that this definition is 'putting words into Schumpeter's mouth' since Schumpeter never defines innovation in these specific, commercial terms. Rather, Godin argues that Schumpeter was more concerned with the recombination of factors of production. Despite Godin's argument otherwise, innovation is still largely conceptualized as the commercialization of research and knowledge in mainstream academic debate. This is evident in the terminology deployed by the journal *Research Policy* – the preeminent innovation studies journal – in its aims and scope:

> *Research Policy* (RP) is a multi-disciplinary journal devoted to analyzing, understanding and effectively responding to the economic, policy, management, organizational, environmental and other challenges posed by innovation, technology, R&D and science. This includes a number of related activities concerned with the creation of knowledge (through research), the diffusion and acquisition of knowledge (e.g. through organizational learning), *and its exploitation in the form of new or improved products, processes or services.* [emphasis added]

Thinking of innovation as commercialization of knowledge does have some benefits, though, in that it helps to differentiate innovation more clearly from research and knowledge. In particular, defining innovation as the commercial application of research and knowledge problematizes some of the assumptions underpinning work in the economics of science (see above). In particular, as a concept innovation provides one analytical tool for understanding how knowledge is not freely acquired, easily accessible, or easily integrated into organizations or wider social systems. Rather, innovation provides one way to think about the geography of research and knowledge, especially the 'stickiness' of place (Markusen 1996). It is evident that innovation – defined as the commercialization of knowledge – is a highly uneven geographical process, in that not all localities or regions produce new products or services at the same rate. As a number of scholars stress, innovation is very much a geographical process as the result of differing local or regional capacities, systemic interdependencies, and locational advantage (e.g. Bracyzk *et al.* 1998; Gertler 2003; Fagerberg 2005). Consequently, it is not surprising that emerging *innovative* sectors like the life sciences are highly concentrated in particular regions.

How do I define the life sciences?

The life sciences loom large in debates about the (geographical) political economy of science, technology and innovation. As an economic sector, it covers a broad set of organizational competencies and capabilities stretching from biomedicine through environmental services to agriculture. It is, in this sense, actually quite difficult to define what we mean by 'life sciences', as its use over the years attests. Generally, it has come to replace 'biotechnology' as a signifier for the research, development and production of services and products based on biological processes and/or knowledge (e.g. CEC 2002). As just mentioned, its intellectual lineage stretches from the previously popular term 'biotechnology' or 'biotech' through to more recent terms like 'bio-economy' or 'bio-based economy' (Staffas *et al.* 2013). It is, therefore, perhaps pertinent to provide a brief history of the life sciences here – conceived in singular terms – in order to ground its importance in debates around regional development and regional policy.

The term 'biotechnology' actually dates back to the First World War when the Hungarian agriculturalist Karl Ereky coined the term to refer to the fattening of pigs (Bud 1993). As such, biotechnology represented a rather loose term to refer to various biological processes (e.g. feeding pigs) as well as knowledge that stretches back centuries, if not millennia (e.g. fermentation). Many commentators, therefore, use 'modern biotechnology' to refer to technoscientific platforms derived from more recent research in genetics, genomics, etc. For example, the UK's seminal 1980 *Spinks Report* – established to provide the UK with an advantage in biotechnology – defined modern biotechnology as 'the application of organisms, systems or processes to manufacturing and service industries' (ACARD *et al.* 1980: 7). Although this definition barely differentiates between modern and old biotechnology, it provided the basis for later definitions that sought to frame biotechnology as an *enabling technology*. In 1993, the House of Lords Select Committee on Science and Technology stated that biotechnology is 'not a single discipline' but rather 'it is a collection of quite different enabling technologies' including recombinant DNA, cell fusion techniques, micro-infections, viruses and fermentation of cell cultures (House of Lords 1993: 14–15).

In the UK at least, these definitions persisted into the late 1990s in various reports by the Department of Trade and Industry (DTI); for example, *Biotechnology Clusters* (DTI 1999a) and *Genome Valley* (DTI 1999b). However, by the 2000s the preferred term shifted to 'bioscience' as evident in the establishment of a Bioscience Innovation and Growth Team (BIGT) (BIGT 2003, 2009). This BIGT sought to include the broadening and emerging technoscientific disciplines catching the eyes of policy-makers and others, including genomics, proteomics, etc., as well as the varied social actors implicated in a national innovation system (e.g. National Health Service, government, Regional Development Agencies, etc.). At the same time, EU policy-makers had shifted to the use of 'life sciences' as a preferred term; for example, the Commission's 2002 *Strategy for Europe on Life Sciences and Biotechnology* defined 'life sciences and biotechnology' broadly as:

> ... one of the most promising frontier technologies for the coming decades. Life sciences and biotechnology are enabling technologies – like information technology, they may be applied for a wide range of purposes for private and public benefits.
>
> (CEC 2002: 5)

This terminological shift inflected subsequent UK government policy as well. In 2009, for example, the Labour Government established an Office for Life Sciences, linking the former DTI and Department for Health, while in 2011 the new Conservative-Liberal Democrat Coalition government launched a *Strategy for UK Life Sciences* (BIUS 2011). The latter defined the life sciences rather simply as 'the application of biology', hardly shifting the underlying conceptual premise from the days of the *Spinks Report* some 30 years earlier.

Although it has become the dominant term, life sciences tends to be used to refer biomedical research and innovation. In the mid-2000s the EC and Organisation for Economic Co-operation and Development (OECD) started to promote another term, namely 'bio-economy' (Birch *et al.* 2010; Mittra 2016). Their definitions rest on the idea of the bio-economy as the:

> ... aggregate set of economic relations in society that use the latent value incumbent in biological products and processes.
>
> (OECD 2006: 1)

Or that:

> The bio-economy is one of the oldest economic sectors known to humanity, and the life sciences and biotechnology are transforming it into one of the newest.
>
> (CEC 2005: 2)

As a new term, the bio-economy is used to cover biomedical, environmental and agricultural research and innovation. Moreover, it is generally associated with the idea of a wholesale, societal transition from fossil fuel dependence to bio-based resource use (Staffas *et al.* 2013). It is, in the context of this book at least, far too broad a definition, especially since much of the empirical research I draw on here has primarily focused on biomedical research and innovation. Thus I prefer to use the term life sciences from now on. I use it in the broad sense highlighted by both UK and European policy documents to cover a range of recent biological research and innovation that has been commercialized in medical applications – covering biopharmaceuticals and medical devices.

Methodological framework and empirical material

This book draws on over a decade of primary and secondary research into the evolving relationship between innovation, regional development and the life sciences. As the political economy of the life sciences has changed over this time, so has my own political-economic understanding of the life sciences. And this has a particular impact on how I present my arguments herein. My concern in writing this book is to avoid being theoretically anachronistic. By this I mean it is difficult not to seek to reinterpret previous research from one's current perspective, drawing on a range of new ideas and empirical material to make a series of new arguments. I do not want to do that in this book. Rather, I think it is important to retain the original conceptual views I held in order to present properly how the research fits within the particular scholarly debates at the time it was undertaken. To this end, my aim is to try to ensure that the research in each chapter is (conceptually) true to the particular time period in which it was originally conceived and carried out. Therefore, I have to take the reader through my intellectual journey from the early 2000s to the present via my research concerns over these years.

It all starts with my doctoral research undertaken in the Department of Planning at Oxford Brookes University between 2002 and 2006. Although I left Oxford Brookes to take up a research fellow job at the University of Glasgow in early 2005, to which I return below, my intellectual development during this time was dominated by debates in regional studies, economic geography, innovation studies, management studies and elsewhere with *industrial clusters* (Porter 1990, 2000), *regional innovation systems* (Bracyzk *et al.* 1998), and *territorial innovation models* (Moulaert and Sekia 2003; Lagendijk 2006). As will become evident later in the book, I found these debates both helpful for focusing my research but increasingly limiting for understanding the life sciences in particular and innovation more generally. The empirical data I draw on here from this time includes: (a) structured interviews with 109 UK-based informants from a range of organizations (e.g. life science firms, policy-makers, service providers, etc.); and (b) secondary data on the UK life sciences sector collected from a variety of sources (Birch 2006a, 2009).

An important part of my intellectual development was moving to the Centre for Public Policy for Regions (CPPR) at the University of Glasgow in early 2005. I ended up working at Glasgow for four years, until early 2009, with economic geographers like Andy Cumbers, Danny MacKinnon, and Andy Pike, as well as my CPPR colleagues Vlad Mykhnenko and Katherine Trebeck. During this time I began to question the spatially localized notions of innovation, developing a programme of research on life sciences using a *global commodity/value chain approach* (Gereffi 1994; Gereffit *et al.* 2005). The ideas and concepts I was introduced to or picked up at CPPR helped me to re-frame how I understood innovation and regional development. I began, for example, to engage with debates about how the local and global interact, exemplified by the work of Bathelt *et al.* (2004), and what this means for innovation governance in regional economies, especially less-favoured regions. In this book, I draw on empirical data collected during this time from two main projects: (a) one small survey of 12 UK life science managers; (b) a larger survey of 72 Scottish life science managers; (c) secondary data on the UK and Scottish life sciences; and (d) in-depth interviews with 32 Scottish-based informants, including life science managers and policy-makers (Birch 2008, 2011; Birch and Cumbers 2010).

In early 2009 I was fortunate enough to get a permanent lecturing job at Strathclyde University in the Department of Geography and Sociology. Again, this led to a change in theoretical direction, especially in terms of my focus on the policy discourses around the life sciences. Even before I began working at Strathclyde I had started researching the role of discourse in the life sciences (e.g. Birch 2006b). Much of this work involved collaborating with Les Levidow and Theo Pappaioannou from the Open University on the policy discourse and policy frameworks engendered by the Commission's *knowledge-based bio-economy* strategy (e.g. Birch *et al.* 2010, 2014). This research was primarily driven by a desire to understand the policy influence of the life sciences sector since one major finding of my PhD (and subsequent research) was that the life sciences had really failed to deliver on its purported innovation promise – i.e. new technological products and services (see Nightingale and Martin 2004; Pisano 2006; Hopkins *et al.* 2007). In this book, I mainly draw on secondary empirical material, specifically policy documents produced by a range of policy-makers from the European Commission through to the UK's now defunct Regional Development Agencies.

Finally, in light of the global financial crisis (GFC) starting in 2007–08, I became interested in the process of financialization (Krippner 2005; Pike and Pollard 2010), and especially the impact of financial practices, processes and knowledges on innovation in the life sciences. This interest blossomed into actual research after I moved to York University in Canada, which happened in mid-2011. After this move I undertook a project in 2012 looking at the global financialization of the UK life sciences, especially relating to the implications and impacts of the GFC on the UK life sciences. From this project, I draw on in-depth interviews with 13 informants from the financial sector who are involved in financing life science businesses (Birch forthcoming).

As this brief intellectual history should indicate, my interest in the life sciences has evolved, as has my interest in the relationship between innovation and regional development. I want to ensure that the subsequent chapters maintain this focus on the specific debates I was engaging with at the particular time I did the research, so that I do not treat the empirical material as directly commensurate or comparable. It was, as with all research, collected with particular theoretical and methodological goals and objectives in mind.

Conclusion and outline of the book

As with any book, it is difficult to provide a pithy summary of my overall objectives. Nevertheless, I will try here. In the rest of the book I want to examine different aspects of the relationship between innovation and regional development using the life sciences as my substantive empirical focus. I started out my research career, way back in 2002, with a fairly open mind about what the life sciences offered society generally and regional economies more specifically. For one, I thought that the life sciences represented an important technoscientific pathway for new healthcare products and services, as well as an important source of high-quality regional employment. As time passed, however, I realized my rose-tinted view of the life sciences was somewhat off. This led me to a number of new research avenues, which I try to bring together here.

I start the book with a discussion of the theoretical debates on the relationship between geography and innovation in **Chapter 2**, especially as they concern the conceptualization of local or regional clusters and regional innovation systems that dominated regional studies and regional policy in the late 1990s and early 2000s (e.g. Porter 2000; Bracyzk *et al.* 1998). In particular, I discuss concepts and debates in regional studies and economic geography on the importance of difference 'territorial innovation models' or TIMs (Moulaert and Sekia 2003; Lagendijk 2006). I do so in order to push forward these analytical frameworks by introducing my own work on *global knowledge-based commodity chains* (Birch 2008, 2011; Birch and Cumbers 2010). The aim in Chapter 2 is to highlight the need to rethink analytically and empirically the relationship between innovation and geography by focusing on questions of economic (and social) value that extend beyond local and linear notions of innovation.

In order to problematize the localized and regional perspectives on innovation, in **Chapter 3** I outline and analyze the particular (and peculiar) geographies of the life sciences sector in the UK. I focus on what I call the *knowledge–space dynamic* in a range of so-called 'biotech clusters' around the UK, examining theoretical arguments about the role and place of tacit knowledge in spatially-bounded innovation systems and processes. In so doing I show that each supposed cluster is different and distinct from the others. In understanding these differences, it is critical to analyze the multi-scalar nature of innovation processes, rather than emphasizing their local or regional dimensions.

Analyzing innovation as a multi-scalar process necessitates adopting a new analytical framework from the ones dominant in regional studies and economic geography at the time of my PhD research; one fruitful approach I found was the global commodity or value chain (GCC/GVC) approach developed by Gary Gereffi (1994) and others (e.g. Gereffi *et al.* 2005). I extend the GCC/GVC approach in **Chapter 4** by applying it to a high-technology sector (i.e. life sciences) in a country and region from the Global North (i.e. Scotland). My goal here is to analyze the positioning of Scotland within the global life sciences sector in order to understand the different forms of innovation governance at play that might enable a 'less-favoured region' like Scotland to capture value from new sectors and upgrade geographical capacities.

In light of the multi-scalar dimensions to innovation, it is necessary to examine a broader set of processes at play in the life sciences. One area that has been neglected in regional development literature, for example, is the role of (policy) discourse; I address this gap in **Chapter 5** by examining the socio-technical imaginaries produced by various policy actors in the European Union, UK national government, and UK regions. My objective in this chapter is to highlight the discursive shaping of (national and regional) institutions that are implicated in the innovation ecosystem (e.g. universities, government, regulations, etc.) (Birch *et al.* 2014).

The wider discussion of the global innovation ecosystem sets up the focus on finance and financialization in **Chapter 6**. More scholars are focusing on how finance shapes research and innovation (e.g. Andersson *et al.* 2010; Lazonick and Tulum 2011; Styhre 2015), but it has been another neglected topic in regional development (Coe *et al.* 2014). As a result of the global financial crisis, however, it has become an increasingly important topic of concern. In this chapter my aim is to analyze the impacts and implications of global finance on the UK life sciences – and, by association, of financialization on life sciences innovation and policy.

In the concluding chapter, **Chapter 7**, I bring all these discussions together to analyze their implications for regional development. I emphasize the need to think about both the economic and social value of the life sciences, especially as it relates to the potential of such emerging sectors to renew declining or less-favoured regions. The chapter will consider how regional policies and strategies have been shaped by a range of processes discussed throughout the book and what this means for a socially-grounded regional policy.

References

ACARD, ABRC, and the Royal Society (1980) *Biotechnology: Report of a Joint Working Party* [aka The Spinks report], London: Her Majesty's Stationery Office.

Andersson, T., Gleadle, P., Haslam, C. and Tsitsianis, N. (2010) Bio-pharma: A financialized business model, *Critical Perspectives in Accounting* 21: 631–641.

Bathelt, H., Malmberg, A. and Maskell, P. (2004) Clusters and knowledge: Local buzz, global pipelines and the process of knowledge creation, *Progress in Human Geography* 28(1): 31–56.

Bell, D. (1973) *The Coming of Post-Industrial Society: A Venture in Social Forecasting*, New York: Basic Books.

BIGT (2003) *Improving National Health, Improving National Wealth*, London: Bioscience Innovation and Growth Team.

BIGT (2009) *The Review and Refresh of Bioscience 2015*, London: Bioscience Innovation Growth Team.

Birch, K. (2006a) *Biotechnology Value Chains as a Case Study of the Knowledge Economy: The Relationship between Knowledge, Space and Technology*, Department of Planning: Oxford Brookes University.

Birch, K. (2006b) The neoliberal underpinnings of the bioeconomy: The ideological discourses and practices of economic competitiveness, *Genomics, Society and Policy* 2(3): 1–15.

Birch, K. (2008) Alliance-driven governance: Applying a global commodity chains approach to the UK biotechnology industry, *Economic Geography* 84(1): 83–103.

Birch, K. (2009) The knowledge–space dynamic in the UK bioeconomy, *Area* 41(3): 273–284.

Birch, K. (2011) 'Weakness' as 'strength' in the Scottish life sciences: Institutional embedding of knowledge-based commodity chains in a less-favoured region, *Growth and Change* 42(1): 71–96.

Birch, K. (forthcoming) Rethinking *value* in the bio-economy: Finance, assetization and the management of value, *Science, Technology and Human Values*.

Birch, K. and Cumbers, A. (2010) Knowledge, space and economic governance: The implications of knowledge-based commodity chains for less-favoured regions, *Environment and Planning A* 42(11): 2581–2601.

Birch, K. and Mykhnenko, V. (2009) Varieties of neoliberalism? Restructuring in large industrially-dependent regions across Western and Eastern Europe, *Journal of Economic Geography* 9(3): 355–380.

Birch, K. and Mykhnenko, V. (2014) Lisbonizing vs. financializing Europe? The Lisbon Strategy and the (un-)making of the European knowledge-based economy, *Environment and Planning C* 32(1): 108–128.

Birch, K., Levidow, L. and Papaioannou, T. (2010) Sustainable Capital? The neoliberalization of nature and knowledge in the European knowledge-based bio-economy, *Sustainability* 2(9): 2898–2918.

Birch, K., Levidow, L. and Papaioannou, T. (2014) Self-fulfilling prophecies of the European knowledge-based bio-economy: The discursive shaping of institutional and policy frameworks in the bio-pharmaceuticals sector, *Journal of the Knowledge Economy* 5(1): 1–18.

BIUS (2011) *Strategy for UK Life Sciences*, London: Department for Business, Innovation and Skills.
Braczyk, H.-J., Cooke, P. and Heidenreich, M. (eds) (1998) *Regional Innovation Systems*, London: UCL Press.
Bud, R. (1993) *The Uses of Life: A History of Biotechnology*, Cambridge, UK: Cambridge University Press.
Castells, M. (1996) *The Rise of the Network Society*, Cambridge, MA; Oxford, UK: Blackwell.
CEC (2002) *Life Sciences and Biotechnology – A Strategy For Europe*, COM(2002) 27 final, Brussels: Commission of the European Communities.
CEC (2005) *New Perspectives on the Knowledge-based Bio-economy: Conference Report*, Brussels: Commission of the European Communities.
CEC (2010) *Communication from The Commission. Europe 2020: A Strategy for Smart, Sustainable and Inclusive Growth*, COM(2010) 2020 final, Brussels: Commission of the European Communities.
Coe, N., Lai, K. and Wojcik, D. (2014) Integrating finance into global production networks, *Regional Studies* 48(5): 761–777.
Cohen, W. and Levinthal, D. (1990) Absorptive capacity: A new perspective on learning and innovation, *Administrative Science Quarterly* 35(1): 128–152.
Dicken (2003) *Global Shift*, London: SAGE.
DTI (1999a) *Biotechnology Clusters Report*, London: Department of Trade and industry.
DTI (1999b) *Genome Valley: The Economic Potential and Strategic Importance of Biotechnology in the UK*, London: Department of Trade and industry.
European Council (2000) *An Agenda of Economic and Social Renewal for Europe* (aka Lisbon Agenda), Brussels: European Council [DOC/00/7].
Fagerberg, J. 2005. Innovation: A guide to the literature, in J. Fagerberg, D. Mowery and R. Nelson (eds) *The Oxford Handbook of Innovation*, Oxford: Oxford University Press, pp.1–26.
Felt, U., Wynne, B., Callon, M., Goncalves, M.E., Jasanoff, S., Jepsen, M., Joly, P.-B., Konopasek, Z., May, S., Neubauer, C., Rip, A., Siune, K., Stirling, A. and Tallacchini, M. (2007) *Science and Governance: Taking European Knowledge Society Seriously*, Brussels: European Commission, EUR 22700.
Florida, R. (2002) *The Rise of the Creative Class*, New York: Perseus Book Group.
Freeman, C. (1974) *The Economics of Industrial Innovation*, Harmondsworth: Penguin.
Fuller, S. (2002) *Knowledge Management Foundations*, Woburn MA: Butterworth-Heinemann.
Gereffi, G. (1994) The organization of buyer-driven global commodity chains: How U.S. retailers shape overseas production networks, in G. Gereffi and M. Korzeniewicz (eds) *Commodity Chains and Global Capitalism*, Westport, CT: Greenwood Press, pp. 95–122.
Gereffi, G., Humphrey, J. and Sturgeon, T. (2005) The governance of global value chains, *Review of International Political Economy* 12: 78–104.
Gertler, M. (2003) Tacit knowledge and the economic geography of context, or The undefinable tacitness of being (there), *Journal of Economic Geography* 3: 75–99.
Gibbons, M., Limoges, C., Nowotny, H., Schwartzman, S., Scott, P. and Trow, M. (1994) *The New Production of Knowledge*, London: SAGE.
Godin, B. (2006a) The knowledge-based economy: Conceptual framework or buzzword?, *Journal of Technology Transfer* 31: 17–30.
Godin, B. (2006b) The linear model of innovation: The historical construction of an analytical framework, *Science, Technology and Human Values* 31(6): 639–667.
Godin, B. (2015) *Innovation Contested*, London: Routledge.
Government of Canada (2014) *Seizing Canada's Moment*, Ottawa: Industry Canada.

HM Treasury (2014) *Our Plan for Growth: Science and Innovation*, London: HM Treasury and Department for Business, Innovation and Skills.
Hopkins, M., Martin, P., Nightingale, P., Kraft, A. and Mahdi, S. (2007) The myth of the biotech revolution: An assessment of technological, clinical and organisational change, *Research Policy* 36: 566–589.
House of Lords (1993) *Select Committee on Science and Technology: Regulation of the United Kingdom Biotechnology Industry and Global Competitiveness*, London: HMSO.
Howells, J. (2002) Tacit knowledge, innovation and economic geography, *Urban Studies* 39: 871–884.
Hudson, R. (1999) The learning economy, the learning firm and the learning region: A sympathetic critique of the limits to learning, *European Urban and Regional Studies* 6(1): 59–72.
Ibert, O. (2007) Towards a geography of knowledge creation: The ambivalences between 'knowledge as an object' and 'knowing in practice', *Regional Studies* 41(1): 103–114.
Jessop, B. (2002) *The Future of the Capitalist State*, Cambridge: Polity Press.
Krippner, G. (2005) The financialization of the American economy, *Socio-Economic Review* 3: 173–208.
Lagendijk, A. (2006) Learning from conceptual flow in regional studies: Framing present debates, unbracketing past debates, *Regional Studies* 40: 385–399.
Lazonick, W. and Tulum, O. (2011) US biopharmaceutical finance and the sustainability of the biotech business model, *Research Policy* 40(9): 1170–1187.
Leadbeater, C. (1999) *Living on Thin Air*, London: Penguin.
Lundvall, B.-A. (ed.) (1992) *National Systems of Innovation*, London: Pinter.
Markusen, A. (1996) Sticky places in slippery space: A typology of industrial districts, *Economic Geography* 72(3): 293–313.
Mirowski, P. (2011) *Science-Mart*, Cambridge, MA: Harvard University Press.
Mittra, J. (2016) *The New Health Bioeconomy: R&D Policy and Innovation for the Twenty-first Century*, Basingstoke: Palgrave Macmillan.
Moulaert, F. and Sekia, F. (2003) Territorial innovation models: A critical survey, *Regional Studies* 37: 289–302.
Nightingale, P. (2003) If Nelson and Winter are only half right about tacit knowledge, which half? A Searlean critique of 'codification', *Industrial and Corporate Change* 12(2): 149–183.
Nightingale, P. and Martin, P. (2004) The myth of the biotech revolution, *Trends in Biotechnology* 22(11): 564–569.
OECD (1996) *The Knowledge-Based Economy*, Paris: Organisation for Economic Co-operation and Development.
OECD (2006) *The Bioeconomy to 2030: Designing a Policy Agenda*, Paris, Organisation for Economic Co-operation and Development.
Pike, A. and Pollard, J. (2010) Economic geographies of financialization, *Economic Geography* 86(1): 29–51.
Piore, M. and Sabel, C. (1984) *The Second Industrial Divide: Possibilities for Prosperity*, New York: Basic Books.
Pisano, G. (2006) *Science Business*, Cambridge, MA: Harvard University Press.
Polanyi, M. (1958) *Personal Knowledge: Towards a Post-Critical Philosophy*, Chicago: University of Chicago Press.
Porter, M. (1990) *The Competitive Advantage of Nations*, London: Macmillan.
Porter, M. (2000) Location, competition, and economic development: Local clusters in a global economy, *Economic Development Quarterly* 14(1): 15–34.
Reich, R. (1991) *The Work of Nations*, New York: Knopf.

Scott, A. (2000a) *Regions and the World Economy*, Oxford: Oxford University Press.

Scott, A. (2000b) Economic geography: The great half-century, in G. Clark, M. Feldman and M. Gertler (eds) *The Oxford Handbook of Economic Geography*, Oxford: Oxford University Press.

Simmie, J. (ed.) (2001) *Innovative Cities*, London: Spon.

Sokol, M. (2004) The 'knowledge economy': A critical view, in P. Cooke and A. Piccaluga (eds) *Regional Economies as Knowledge Laboratories*, Cheltenham: Edward Elgar, pp. 216–231.

Staffas, L., Gustavsson, M. and McCormick, K. (2013) Strategies and policies for the bioeconomy and bio-based economy: An analysis of official national approaches, *Sustainability* 5(6): 2751–2769.

Storper, M. (1997) *The Regional World: Territorial Development in a Global Economy*, London: Guilford Press.

Styhre, A. (2015) *Financing Life Science Innovation: Venture Capital, Corporate Governance and Commercialization*, Basingstoke: Palgrave Macmillan.

Tyfield, D. (2012a) *The Economics of Science*, London: Routledge.

Tyfield, D. (2012b) A cultural political economy of research and innovation in an age of crisis, *Minerva* 50(2): 149–167.

von Hippel, E. (2005) *Democratizing Innovation*, Cambridge, MA: MIT Press.

Williams, A. (1992) *The Western European Economy: A Geography of Post-war Development*, London: Routledge.

2 Innovation, clusters and knowledge-based commodity chains

Introduction

As economics has developed through the 20th century, it has paid closer and closer attention to the role of knowledge in economic growth and development. Concomitantly, academics, commentators and policy-makers have all sought to enrol knowledge in their visions of the future, as mentioned in the book's introduction. Epithets like 'knowledge-based', 'knowledge-driven', 'information', 'post-industrial' and so on have all been used to describe the societal transition from manufacturing to service employment (Cooke and Leydesdorff 2006; Godin 2006a). I will come back to the intellectual history and lineage of these concepts and theories shortly. In the meantime, suffice to say that they have been incredibly influential with a number of recent and important policy agendas stressing the important role of knowledge to our societies and economies. For example, the OECD (1996) and EC (2000) have both promoted the emergence of a *knowledge-based economy* as both a social reality and policy priority in face of global economic imperatives. Other policy actors have made similar claims, from global institutions through national governments to regional policy-makers. All of these social actors stress the need to up knowledge intensity across industries, invest more in knowledge producing activities, and support the production and consumption of knowledge products (see Powell and Snellman 2004). And this is where the life sciences sector comes into the picture.

The life sciences sector represents one of the emblematic industries of the knowledge-based economy in these policy discourses and academic debates; for example, it relies on new scientific knowledge, it entails high levels of scientific education, and it promises high-tech products and services. Despite its emblematic status, the life sciences sector is rather disappointing when it comes to on-the-ground innovation (Nightingale and Martin 2004; Pisano 2006). Notwithstanding its patchy record as a symbol of high-tech futures, the life sciences has been centrally placed in many national and regional policy visions and strategies, especially in the UK. In part, the life sciences sector has assumed a discursive place and role in promoting a particular approach to economic development generally and regional development more specifically. It is associated, for example, with a series of technoscientific developments over the last 40 years or so including recombinant

DNA (rDNA), cell fusion, protein engineering, genomics, stem cells and others (Woiceshyn 1995; Acharya 1999). Early biotech businesses like Genentech – founded in 1976 as a spin-off from the work of Stanley Cohen, Herbert Boyer and colleagues at Stanford University and the University of California – became one of the most profitable stock market flotations of all time in 1980 (Dutfield 2003). It is hardly surprising that many policy-makers have rushed to support the life sciences as a panacea to declining manufacturing competitiveness and employment.

Whether or not the life sciences represent a political-economic answer to the ongoing transformation of our economies is somewhat beside the point. Considerable effort, money and time has been invested in promoting and supporting the life sciences, from institutional changes (e.g. changing intellectual property laws) through financial investment (e.g. venture capital) to public awareness raising (e.g. consumer acceptance) (Pisano 2006; Birch 2007). The life sciences sector is, in light of these changes, an important sector to consider. It is also useful to analyze because it provides a series of insights relevant to other emerging high-technology sectors (e.g. nanotech). How to understand the life sciences conceptually is, therefore, an important task to undertake, especially in light of policy agendas and strategies based on recommendations derived from specific theories (e.g. clusters).

My aim in this chapter is to ground the rest of the book in the major theoretical debates on the relationship between innovation and geography. I start by providing a general overview of the history of scholarly debates about the relationship between knowledge, innovation and economic growth, especially as they relate to the economics of science and innovation. I then turn to later debates around local and regional innovation systems and clusters that dominated regional studies and economic geography in the late 1990s and early 2000s. I do this in order to frame my research agenda in later chapters, especially in Chapter 3 where I analyze the geography of the UK life sciences sector. I then turn to the specific literature on the life sciences, before ending with my theoretical framework for later chapters in the book. This framework is based on the concept of *knowledge-based commodity chains* which represents my attempt to integrate multi-scalar processes specific to knowledge-based sectors.

Knowledge, innovation and political economy

The role of knowledge and innovation in economic development and growth has been highlighted since the early-to-mid 20th century by a number of economists and others. Here I refer to two sets of argument based on concepts drawn from mainstream economics and from innovation studies. Before discussing these two sets of ideas, it is worth noting that the political economy of research and innovation moves through different phases. According to Mirowski and Sent (2008), for example, there have been three regimes of science organization in 20th century America: (1) a Chandlerian regime (1890–Second World War) in which the emergence of research laboratories was tied to the rise of large corporations; (2) a Cold War regime (Second World War–1980) in which research and innovation were tied to national military objectives; and (3) a globalized privatization regime characterized by knowledge enclosure, privatization and commodification. The reason I mention this now is to emphasize the relationship between knowledge production, innovation and broader political-economic trends.

First, some of the earliest exponents of the view that knowledge is a vital factor of production were mainstream economists like Robert Solow (1956). He argued that 'technical change' helped explain the 'residual' left over when trying to explain economic growth in terms of labour and capital contributions (see Nelson and Winter 1982). However, this residual is problematic because it covers a range of possible changes, including organizational, epistemic and educational, that could impact economic development (Coombs *et al.* 1987). As a result of these ideas and other influences – like Schumpeter discussed below and Vannevar Bush's 1945 *Science, the Endless Frontiers* report – economists turned to knowledge and innovation as key objects of their own attention. For example, Godin (2006b) provides an account of the development of the linear model of innovation resulting from the interest of economists in the impact of scientific research on the economy. As mentioned in the Introduction, this linear model has been criticized subsequently (e.g. Kline and Rosenberg 1986), but it remains an influential framing device in much science and innovation policy (e.g. Government of Canada 2014; HM Treasury 2014).

The political-economic importance of knowledge and innovation to the economy has spread through various wider debates about the shift to or emergence of *knowledge economies* or *societies*. For example, one of the first people to identify the emergence of a knowledge economy was Fritz Machlup (1962); he argued that an increasing proportion of US GNP belonged to 'knowledge industries', defined as information centred industries which make an impression on people's minds (Brint 2001). While Machlup's claims were contested, he was merely the start of a trend of scholars proclaiming the start of a new, knowledge-based age. Others, like Daniel Bell (1973) who posited a shift from industrial to 'post-industrial' society, argued that the US economy was increasingly dependent upon the service sector and science-driven, high-tech industries. Since the 1960s and 1970s numerous thinkers have argued that we are witnessing such a shift, whether to an 'information society', 'knowledge society', 'new economy', 'network society', 'weightless economy', or something similar (for reviews see, Fuller 2001; Thompson 2004; Sokol 2004; Godin 2006a). It is worth noting that many of these breathless paeans to knowledge futures were written before the Dot.Com crash of 2000. That being said, concepts like the *knowledge-based economy* became influential policy narratives, espoused by the OECD (1996) and European Council (2000) – see Birch and Mykhnenko (2014).

Second, another intellectual tradition has emerged alongside mainstream economics which is generally referred to as innovation studies. It draws on work by a range of heterodox economists including people like Joseph Schumpeter. The Austrian economist Joseph Schumpeter (1939, 1942) is frequently ascribed the role of introducing innovation as *the* key explanation for capitalism's continuing dynamism and survival. In particular, he is known for introducing the idea that there are regular 'gales [of] creative destruction' that wipe out existing, incumbent industries as the result of the emergence of new industries based on new technologies. Creative destruction is based on the idea that economies are usually based on a dominant technological innovation (e.g. steam engine) that gradually spreads throughout society before being superseded by another technological innovation (e.g. steel). Entrepreneurship is critical to this process as individuals (early Schumpeter) and businesses (later Schumpeter) research, develop and spread new innovations (see Simmie 2003). As such, Schumpeter and his later adherents contend that capitalism is evolutionary rather than a static equilibrium (see Nelson and Winter 1982).

Schumpeter's work has inspired considerable research in the area of innovation studies (Fagerberg and Verspagen 2009). Drawing on the early work of scholars like Schumpeter, Christopher Freeman, Kenneth Arrow, and Nelson and Winter, innovation studies focuses on the 'commercialization of technological inventions' (Godin 2012: 398). As a recent edited collection, *Innovation Studies*, states:

> ... to learn more about how society can benefit from innovation, one also needs to understand innovation processes in firms and how these interact with broader social, institutional, and political factors.
>
> (Fagerberg *et al.* 2013: 1)

In these terms, innovation is defined in firm-centric terms, primarily as the commercial introduction of new processes, products and services, especially if they are characterized as 'technological' (i.e. artefacts). Some, especially Godin (2012, 2014, 2015), have criticized this narrow focus, highlighting that Schumpeter, for example, included 'new commodities, new methods, new forms of organization, new sources of supply, and new markets' in his definition of innovation (Godin 2012: 406). That notwithstanding, innovation studies identifies a series of important factors in the analysis of innovation; these include: firms, users, learning, capabilities, collaboration and collective action, role of the state, and systemic relations and ecosystems. It is the last of these that informs the following theoretical discussion on the geography of innovation.

Innovation, geography and regional development

As debates about innovation have developed, the idea that innovation is best thought of as a systemic process has become increasingly important. Fagerberg (2005: 3) argues that this systems perspective was first introduced in Freeman *et al.*'s (1982) book, *Unemployment and Technical Innovation*. As a systemic process, then, any understanding of innovation necessitates the analysis of both organizational sites (e.g. firms) *and* specific geographical locations (e.g. regions) (Fagerberg 2005). The concept of systems of innovation has subsequently been theorized in national systems terms (e.g. Lundvall 1992) and regional systems terms (e.g. Cooke 1998, 2004d; Braczyk *et al.* 1998). In both cases, a range of constituent system elements have been identified as contributing to innovation, including: formal and informal institutions such as regulations or trust; political process and government support like infrastructure spending; public organizations like universities or research centres; public and private financial actors like business angels and venture capitalists; labour market characteristics like skills and training levels; and so on. This systemic view of innovation has been incredibly influential in debates on the importance of innovation to regional development, primarily because of the uneven geographical distribution of many of these elements in different localities and regions.

Empirically, it is evident that innovation is unevenly spread across and within countries; for example, sites of research (e.g. universities) are located in certain places, venture capital tends to be concentrated in certain regions and not others, and new products tend to be developed in particular places (e.g. Silicon Valley). While these geographical differences and unevenness are dynamic (Hudson 2005), in that they change over time, they still have a direct impact on the capacity of different regions to pursue specific forms of economic development (Feldman 1999, 2000). Such uneven development has a long intellectual history in regional studies and economic geography, going back to the work of geographers like David Harvey (1999[1982]), Neil Smith (2008[1984]) and Doreen Massey (1995[1984]).

Subsequent work has focused on the (regional) geography of innovation, especially scholars writing in the late 1990s and early 2000s on the 'new regionalism'. A number of reviews of this literature are available, including: critical accounts by Lovering (1999); discussions by MacLeod (2001) and MacKinnon *et al.* (2002); and overviews of 'territorial innovation models' (TIMs) by Moulaert and Sekia (2003), Lagendijk (2006) and Pike *et al.* (2006). In the rest of this section I outline the main theoretical trends in these debates. In his review, Lagendijk (2006) argues that there are three phases to the development of conceptual perspectives on the geography of innovation, or TIMs. They start with an early 'structuralist-organizational' phase based on work from the likes of the California School; then they move into a 'social-institutional' phase with work on 'institutional thickness' (Amin and Thrift 1992); and they end with the 'cognitive' phase focusing on communities of practice. For my purposes here, I define these phases as *functional*, *relational* and *associational* theories of spatial innovation systems. By this I mean that functional theories focus on material linkages in space; relational theories focus on social and institutional linkages; and associational theories focus on cognitive and practice-based linkages.

First, the *functional* theories build upon work in economics on transaction costs, the evolution of technological paradigms and trajectories, and French Regulation theory. The early work of the California School in the 1980s is one example of this research agenda. Scholars focused on the importance of transaction costs and flexible production enabled by existing Fordist systems of accumulation to the production of 'new industrial spaces' (NIS) (e.g. Scott 1989; Storper and Walker 1989). The prime example of an NIS was Silicon Valley, California (Scott 1989), where production was distributed across numerous firms such that external agglomeration economies come to represent a significant regional advantage to firms in those places (Lagendijk 2006). This work provided an impetus to the increasing conceptualization of regions as key sites of economic and social activity in post-Fordist economies (Storper 1995; Storper and Scott 1995). As Allen Scott (1998: 387) argued, agglomeration economies and the spatial concentration of production produced 'a widening of the social division of labor', which thereby widened knowledge and innovation capacities. As Moulaert and Sekia (2003) note, similar ideas are found in other concepts like 'flexible specialization' and 'industrial districts'. All this research emphasized the distributed nature of production across a range of localized small firms, each undertaking a distinct part in the production process and trading with one another. Regional advantage was premised on the geographical co-location of all parts of the supply chain in one place, enabling firms along the supply chain easy access to each other and the development of inter-dependent learning.

Second, some functional theorists, like Michael Storper (1995), began to incorporate 'untraded interdependencies' into their perspective, highlighting the importance of *relational* concepts to understandings of the geography of innovation. A range of relational theories stress the need to understand the impact of social and institutional characteristics of regions on the innovation process. These theories built upon work in economic sociology on the embeddedness of economies in society (e.g. Granovetter 1985) as well as the systemic nature of innovation from the innovation studies literature (see above) (Fagerberg 2005). Conceptualizing innovation in systemic and embedded terms meant these relational theories emphasized the importance of collective and interactive learning processes in understanding the geography of innovation (Morgan 1997; Malecki 2000). There are a number of key relational theories worth highlighting at this point, including (but not limited to): Michael Porter's (1990, 2000) concept of the geographical cluster, whether national or regional; Amin and Thrift's (1992, 1994) notion of 'institutional thickness' to conceptualize the institutions (e.g. trade associations, regional government, universities, etc.) surrounding firms in particular places; Camagni (1995) and others' concept of 'innovative milieu' to conceptualize the surrounding social, political, cultural, etc. factors that play a role in innovation; Kevin Morgan's (1997) concept of the 'learning region' as a way to capture the collective innovation process; and Philip Cooke (1998, 2001a, 2004d) and others' concept of 'regional innovation systems', which focuses on the systemic interaction between institutions, firms and other social actors in particular places.

Many of these relational theories are underpinned by the same concerns with the role of regional institutions, collective local action and interaction, and local culture and trust in engendering innovation. Most also emphasize the central explanatory role of spatial concentration in innovation processes; that is, spatial concentration stimulates innovation through the co-location of complementary assets and capacities that are easier to access and integrate as the result of shared local institutions, culture and trust. Moreover, in emphasizing the spatial dimensions of these processes these theories tend to assume that innovation is driven by the concentration or 'clustering' of research, production and other activities.

One important theory in this regard is the geographical cluster idea drawn from the work of management guru Michael Porter (1990, 2000). Originally a national scale concept, Porter subsequently adapted it to other geographical scales, including the regional. Although it has been criticized by a number of people (e.g. Martin and Sunley 2003), it has proven to be a very durable and popular policy strategy (e.g. Rosiello 2004).

Finally, a final range of explanations for the geography of innovation come from what I call *associational* theories, and what Lagendijk (2006) calls 'cognitive' TIMs. Of particular importance here are the role of knowledge or cognitive communities, cognitive capacities, and cognitive interaction. Although Cooke and Morgan (1998) wrote a book about *The Associational Economy*, my use of the term goes beyond their use. When I use the term, I want to emphasize the specificities of knowledge and learning themselves (Moodysson *et al.* 2008). As several non-geographers point out, interaction is a learning activity itself (e.g. von Hippel 1988, 1994), meaning that geographical proximity does not define the act of interaction itself (Boschma 2005). What matters, in this sense, is the cognitive capacity of social actors to find, acquire and integrate knowledge from wherever, which is facilitated by sharing cognitive values, assumptions, etc. One important concept is the idea of 'communities of practice', which represent groups of social actors who cooperate across organizations and regions because of their shared ways of thinking and knowing (Henry and Pinch 2000; Pinch *et al.* 2003; Bathelt and Gluckler 2005; Asheim and Gertler 2005). An important contribution to these theories is the work of Bathelt *et al.* (2004) on 'local buzz' and 'global pipelines' because it combines the localized social interaction (e.g. buzz) with global social interaction (e.g. pipelines). As I have argued elsewhere (Birch 2012), this local–global dynamic is a central aspect of the life sciences in that the sector is characterized by *both* localized, tacit knowledge production (e.g. firm-level routines) and globalized, abstract knowledge production (e.g. international regulations). This work of Bathelt *et al.* (2004) and others on communities of practice has helped to broaden debates on the geography of innovation from the region-focus of earlier theories, opening up room for multi-scalar perspectives on the relationship between regional development and innovation.

Innovation, geography and the life sciences

Before turning to my theoretical framework on the multi-scalar dynamics of innovation that informs the rest of the book, I want to provide a run through of the relationship between geographical (or otherwise) theories of innovation and the life sciences. I come back to these in the next chapter when I present the concept of a *knowledge–space dynamic* in the UK life sciences. So all I am going to do here is outline the debates around innovation in the life sciences sector. Again, there is an intellectual history to this literature that stretches back to the early 1990s. Generally I split the debate between (1) research focusing on knowledge processes and (2) research focusing on spatial processes.

Knowledge-based theories of innovation in the life sciences

First, the earliest literature comes from strategic management, economic sociology and innovation studies. It was largely focused on things like the performance of alliances and collaboration between new dedicated biotechnology firms (or DBFs) and incumbent pharmaceutical firms, networks of academic and industry actors, and differences between national systems of (biotech) innovation.

The strategic management literature did not really address the issue of territorial innovation processes, but rather concentrated on the importance of alliance and collaboration for DBFs (e.g. Hamilton *et al.* 1990; Chakarabarti and Weisenfeld 1991; Dodgson 1991; della Valle and Gambardella 1993; Woiceshyn 1995; Deeds and Hill 1996; see Bagchi-Sen *et al.* 2001 and Hall and Bagchi-Sen 2001 for reviews). Such research puts a particular stress on the important role and effect of complementarity in such linkages (Dodgson 1991; Deeds and Hill 1996), especially between newly emerging DBFs and existing large pharmaceutical companies, and their impact on the organization of biomedical research (see Senker 2005 for reviews). Inter-organizational collaboration not only provides the means to access capabilities that biotech firms do not have, such as marketing and manufacturing (Woiceshyn 1995), it also means that biotech firms can avoid some of the uncertainty and risks involved in innovation (Chakrabarti and Weisenfeld 1991). As a consequence, della Valle and Gambardella (1993) note that there has been a shift towards more 'open innovation'; an argument that is reiterated later by Cooke (2006). Thus there is an inevitable concern with whether such inter-organizational collaborations lead to integration between different organizations (e.g. Prevezer and Toker 1996) both 'upstream' (e.g. universities) and 'downstream' (e.g. large pharmaceutical firms) from biotech firms (see Hamilton *et al.* 1990; Chakrabarti and Weisenfeld 1991; Dodgson 1991; Greis *et al.* 1995; Deeds and Hill 1996).

This concern with integration helps to explain the sociological interest in the importance of networks in the biotech industry, and draws especially on concepts from 'new economic sociology' (Granovetter 1985) and transaction costs. Such research is exemplified by the work of Walter Powell and colleagues (e.g. Powell *et al.* 1996; Powell 1998; Owen-Smith *et al.* 2002; Owen-Smith and Powell 2003; Powell *et al.* 2002; Powell *et al.* 2004). For example, their research claimed that inter-organizational collaboration networks constitute the 'locus of innovation' (Powell *et al.* 1996). From this perspective, they are most concerned with how such networks have developed and what effects this expansion has had on learning and innovation not only in terms of R&D (Powell *et al.* 2002); however, there is also a concern with the relationship between 'science and capital' (Powell *et al.* 2002). The focus on learning through collaboration (e.g. Powell 1998) helps to distinguish this literature from the early research in strategic management, although they still both share the emphasis on inter-organizational collaboration in the biomedical industry. Such concerns obviously lead to an interest in the differences between countries (e.g. Owen-Smith *et al.* 2002) that connects with work in innovation studies.

The strategic management literature did draw on evolutionary economics and Schumpeterian concepts of innovation (e.g. Hamilton *et al.* 1990), but its explicit focus is rather different from innovation studies. Although research in innovation studies also highlights the important role of inter-organizational collaboration and alliances, it draws on a *systemic* perspective connecting specific national institutions with firm-level competitiveness. For example, a number of authors have argued that biotech firms benefit from advantages provided by their location in specific countries such as the USA, whilst other locations (e.g. Europe) provide a less hospitable environment (see Walsh *et al.* 1995; Acharya *et al.* 1998; Saviotti 1998; Saviotti *et al.* 1998; Senker 1998, 2004, 2005; Acharya 1999). This research builds on a systems of innovation approach inspired by Schumpeterian and evolutionary economics (see Nelson and Winter 1982; Dosi 1988), as well as historical approaches to understanding technological change (e.g. Rosenberg 1976; Freeman 1982). It is primarily centred on the firm and specific technological regimes, in contrast to the network perspective discussed above, concentrating on the role of DBFs in innovation systems by stressing the importance of sectoral and institutional conditions (e.g. Acharya *et al.* 1998; Saviotti 1998; Acharya 1999). These conditions include university–industry relations and the commercialization of biotechnology in small and emerging biotech firms (Sharp and Senker 1999), as well as the impact of national institutions on regulations and intellectual property (IP) (e.g. Senker 2004).

Spatial-based theories of innovation in the life sciences

Geographical perspectives on innovation in the life sciences have moved through a number of strands of research. This includes comparative political economy perspectives; 'new economic geography' (NEG) work on knowledge spillovers; and regional studies and economic geography *proper* on (1) local clusters and regional innovation systems and (2) local–global interaction.

The interest in the political economy of biotechnology has a long history (e.g. Yoxen 1981; Loeppky 2004, 2005; Birch 2006; Benner and Lofgren 2007) and has a number of connections with the work on innovation systems discussed above, particularly when it is focused on national systems of innovation (e.g. Walsh *et al.* 1995). There was some interest in applying the varieties of capitalism (VOC) approach developed by Hall and Soskice (2001) to the life sciences, not only at the national scale (e.g. Kettler and Casper 2000; Casper and Kettler 2001), but also regionally (Casper and Karamanos 2003; Casper and Murray 2004). Such comparative approaches help illustrate the extent to which institutional changes at the national scale provide the necessary environment in which biotechnology can develop. A number of authors have looked at these types of differences between countries when it comes to the life sciences. For example, Niosi and Bas (2004) consider the Canadian government's attempts to promote biotechnology; Walsh *et al.* (1995) examine the national institutional environment in France, Britain and Canada; and Senker (2004) considers national policies across Europe. In this research there is an emphasis on public sector expenditure, financial capital availability and regulatory constraints in different countries.

The initial impetus behind more regional scale research on life sciences innovation was driven by new economic geography (NEG). In particular, the concept of knowledge spillovers was used to explain the existence of increasing returns and agglomeration economies. Research in this area has concentrated on knowledge spillovers from academic research (e.g. Acs *et al.* 1991) as well as industrial R&D, especially in small firms (see Feldman 1999 for a review). Such research can be seen as a resurgence of interest in the *location* of innovation in neoclassical economics. The specific NEG focus on life sciences innovation draws on these insights, although mostly without considering the evolutionary perspective of earlier innovation studies approaches. There has been a particular interest in the relationship between the location of scientists and life sciences (e.g. Audretsch and Stephan 1996; Prevezer 1997), illustrating the relationship between biotechnology firms and academic knowledge. Furthermore, there are important links with the research in economic sociology on network collaborations, especially in the work on human capital and the importance of 'star scientists' to life sciences firms (e.g. Zucker *et al.* 1998; Zucker *et al.* 2002). Here scholars argue that there is a strong connection between human capital (i.e. 'star scientists') and the location of new life sciences firms (Zucker *et al.* 1998) resulting from the importance of tacit knowledge embodied in specific individuals and their networks (Zucker *et al.* 2002).

Interest in the spatial dimensions of life science innovation also emerged in regional studies and (old) economic geography, especially around the concepts of *clusters* and *regional innovation systems*. Despite policy popularity (e.g. DTI 1999a, 1999b), there is surprisingly little research in regional studies and economic geography on the life sciences that specifically adopts Porter's framework (see Cooke 2001, 2003b; Leibovitz 2004; Ryan and Phillips 2004; Asheim and Coenen 2006). Even where scholars studying biotechnology have adopted Porter's approach they are often critical of it (e.g. Leibovitz 2004), as are other academics more generally (e.g. Martin and Sunley 2003). Instead the word 'cluster' has become a generic term to represent any form of geographical concentration of the life sciences that exhibits certain characteristics. These include:

- Concentrations of life sciences firms, which are usually small or medium sized enterprises (SMEs);
- Collaborations with complementary 'upstream' (e.g. universities) and sometimes 'downstream' (e.g. large pharmaceutical firms) organizations;
- Linkages to specialized service organizations like lawyers, business consultants, and accountants;
- 'Cluster' identity through the trade associations and networking organizations;
- Local and national government policy to promote and encourage territorial approaches to economic development.

In contrast to the cluster approach, the regional innovation systems (RIS) perspective has proven particularly popular in regional studies and economic geography when it comes to the life sciences (e.g. Zeller 2001; Ossenbrugge and Zeller 2002; Asheim and Coenen 2006). The RIS approach weds evolutionary, firm and systems theories together. The most prolific scholar in this field is Phil Cooke, whose work encapsulates many of the diverse conceptual strands (e.g. Cooke 2001, 2002, 2003a, 2004a, 2004b, 2004c, 2004d, 2006). Although the RIS concept represents life sciences concentrations in a similar fashion to the characteristics highlighted above, it situates this within specific regional infrastructure and knowledge capabilities (Cooke and Leydesdorff 2006) that differentiates it from the sectoral focus of cluster theory (Asheim and Coenen 2006). Thus it consists of more than just a dense network of functionally similar firms; instead it stretches across numerous place-specific organizations and institutions (e.g. life science firms, business services, public research, education, labour movement, etc.). Furthermore, the RIS approach emphasizes the importance of tacit knowledge and social capital (e.g. trust) that are embedded in these place-specific systems (e.g. Breschi *et al.* 2001; Zeller 2001; Feldman and Francis 2003; Ryan and Phillips 2004), rather than assuming that innovation depends upon internal firm capabilities and knowledge spillovers.

Finally, an important thing to note here is that the focus on the regional scale has perhaps obfuscated some of the extra-local linkages that are crucial to the life sciences, both in relation to knowledge interactions and the innovation process. A concern with extra-regional connections and interactions can be seen as a consequence of the relational turn in economic geography (e.g. Bathelt *et al.* 2004). Whereas there is little empirical evidence to support the contention of localized knowledge linkages (see Malmberg and Maskell 2002; Malmberg 2003; Malmberg and Power 2005), there is ample evidence that extra-local and particularly global ties are central to the success of the life sciences (e.g. Lawton-Smith *et al.* 2000; Walcott 2001; Leibovitz 2004; Ryan and Phillips 2004; Zeller 2004). Consequently, economic geography and regional studies research has increasingly emphasized the importance of engaging with local–global interactions and multi-scalar processes; examples of this work include concepts like ‘local nodes in global networks’ (Coenen *et al.* 2004; Gertler and Levitte 2005), ‘bioscience megacentres’ (Cooke 2002, 2004a, 2004c, 2006), and ‘knowledge-based commodity chains’ to which I turn next (Birch 2008, 2011; Birch and Cumbers 2010).

Knowledge-based commodity chains: going beyond clusters

Much of the regional studies and economic geography literature on life sciences innovation has, up to the early 2000s, focused on local or regional linkages, institutions, and processes. However, although the stylized representation of innovation ‘clusters’ was widely publicized, especially in the life sciences, it is important to note that extra-local connections beyond localized innovation systems (i.e. the ‘cluster’) are as important for firms as local linkages. Extra-local linkages provide important complementary capabilities like manufacturing (Gray and Parker

1998) and the stimuli of international markets and global knowledge interactions (Simmie 2004). As a result, subsequent research on the life sciences focused on the 'nodes of excellence' or 'megacentres' in the global network of biotechnology capabilities (e.g. Coenen *et al.* 2004; Cooke 2004a, 2004c; Cooke and Leydesdorff 2006). However, despite an important consideration of global connections, this research did not address the multi-scalar (local, national and international) linkages in biotechnology innovation processes.

These later debates about biotechnology nodes of excellence or megacentres still emphasize the concentration of whole 'value chains' in particular locations (Coenen *et al.* 2004; Cooke 2004a, 2004c; Cooke and Leydesdorff 2006; Zeller 2004), rather than on multi-scale processes and networks. For example, Cooke (2004c) highlights the importance of 'regional knowledge capabilities' and the shift to diversification as a cluster develops. In other work, there has been a greater emphasis on the issue of proximity and exactly what types of proximity prove central in biotechnology clusters (e.g. Coenen *et al.* 2004; Zeller 2004; see Boschma 2005 for more general discussion), particularly the importance of 'functional proximity' (accessibility) (Coenen *et al.* 2004).

Although this later work was more sophisticated, it still emphasized localized processes. In particular, it promotes the idea of studying the interactions between firms and other organizations at the local or regional scale. In contrast, research has shown that it is important to explore the concentration *and* dispersal of innovation processes across multiple scales (Malmberg 2003; Malmberg and Power 2005), primarily because local external economies from concentration produce both advantages and disadvantages for firms. In their overview of the more general literature on clusters, Malmberg and Power (2005: 425) claimed that there is actually little evidence that firms interact or collaborate more with other local firms and concluded that 'collaborative interaction with similar and related firms in the localized cluster does not come out as a major knowledge creating mechanism.' A number of other scholars have stressed the multi-scale dimensions of innovation, especially the role of global networks and connections (e.g. Bunnell and Coe 2001), which makes it important to consider such extra-local linkages in the life sciences sector.

From the early 2000s and to this end, the literature on the geography of innovation incorporated a range of ideas derived from political economy and economic sociology including approaches like *global commodity chains* (GCC) analysis. This approach has its origins in research of Gary Gereffi (1994, 1996) who focused on the political-economic governance of inter-firm relations and the benefits of technological upgrading this could have for 'less developed countries' (or regions). Gereffi identified two governance models – producer-driven and buyer-driven – each focused on different manufacturing sectors – consumer durables (e.g., automobiles) and consumer non-durables (e.g., apparel), respectively. According to Gereffi (1996) and later writers in this tradition, the main focus of the GCC approach provides a number of conceptual benefits (see Table 2.1). However, there are also several weaknesses identified by a number of human geographers, which I come back to below.

Table 2.1 Analytical benefits of GCC approach

GCC foci	*Benefits*
Organizational aspects of chains and linkages between different economic networks	Can analyze interplay between different institutional systems, rather than assume one system represents dominant form
Cross-national nature of organizations	Can analyze linkages across local, national and global scales, rather be limited to local interactions
Spatial dispersal of governance	Can analyze governance and power relations between actors in chain
Inter- and intra-sectoral variations	Can analyze commodity, rather than sector, firm or location, which enables analysis of variations within and among different sectors

Source: adapted from Birch (2008).

Researchers adopting the GGC approach have made attempts to link the dispersed organization of GCCs with the localized and concentrated organization and governance of production. Much of this work has adopted an updated version of the GCC, which Gereffi *et al.* (2005) define as a global value chain (GVC) approach. Again, the GVC approach has weaknesses, such as a narrow emphasis on the 'internal logics of sectors', rather than on the external linkages among different sectors, organizations, and institutions (see Bair 2005: 164). The concentration on inter-firm relations ignores, analytically-speaking, the differences in regional and national economies and leaves out the relevance of place and the geographic embedding of institutions, path dependence, and regulation. Moreover, the GCC approach has relatively little to say about economies where knowledge and other intangible assets represent an increasingly dominant – if not already dominant (Birch 2015) – proportion of economic activity. Knowledge is not only now a dominant part of many economies and a crucial asset for many firms, small and large, it is also increasingly commodified, packaged, marketed and sold as a commodity, just like more tangible products (see Jessop 2000; May 2000).

When human geographers started working in this area in the early 2000s, they criticized the spatial disembedding implicit in the GCC characterization of production, organizational and governance activities as a linear process, rather than one that is geographically and historically bounded (e.g. Henderson *et al.* 2002; Coe *et al.* 2004; Hess and Yeung 2006). These geographers preferred to conceptualize this spatially sensitive approach as global production networks (GPNs). When it comes to the geography of innovation, the work of Henderson *et al.* (2002), Coe *et al.* (2004), and others on GPNs can help to integrate the differences in innovation processes and governance across different sectoral, organizational and institutional networks at different spatial scales, which are embedded in processes of geographic concentration and dispersal (see Ernst and Kim 2002). The GPN framework is both spatial, concerning the differences between actors at different scales, and relational, addressing the relationships between these actors within and across the different scales.

For my purposes, however, the GPN approach has at least one major drawback, which necessitates a fairly simple solution. A GPN can include almost anything as a constituent element in the network, from labour relations within a firm to global governance treaties between nations. It is, in this sense, too vast to apply practically in any single research project, or even across several. My solution has been to retain the GCC focus on a technological artefact (or 'commodity') as the unit of analysis – rather than focus on a firm or region – and extend the geographical analytical lens accordingly. Furthermore, I have focused more on innovation processes than production or consumption and, as a result, I seek to analyze the governance, discursive and financial processes of innovation as they play out along what I call *global knowledge-based commodity chains* that are dependent on alliance-driven forms of organizational interaction (cf. producer or consumer driven) (Birch 2008, 2011; Birch and Cumbers 2010). It is important to consider the geographies of these knowledge-based commodity chains in order to analyze innovation governance in the life sciences; that is, knowledge creation, transfer, absorption and commodification. This process crosses a number of different organizational (e.g. universities), sectoral (e.g. life sciences), and spatial (e.g. clusters) networks, themselves embedded in different institutional contexts which are, in turn, constituted as much by their relation to other geographies (e.g. their extra-local linkages) as by their particular, locational characteristics.

Knowledge-based commodity chains (KBCC) are constituted by different processes all of which necessitate access to, as well as the purchase and absorption of, different types of knowledge from scientific discoveries through to marketing reports. Much of this knowledge is embedded in certain places as a consequence of spatial innovation systems that are underpinned by regionally-, nationally- and globally-specific institutions (Cooke, 2004a). For example, universities constitute part of the regional education infrastructure in which new science and technology are developed; healthcare systems constitute part of the national market in which firms have to compete; and trade and standards organisations (e.g. WTO, ISO, etc.) constitute part of the global regulatory framework in which firms operate. Consequently, it is theoretically and methodologically vital to go beyond the emphasis on localized knowledge interaction and learning that characterize much of the existing literature on innovation in knowledge-based sectors like the life sciences (Birch 2012).

Using the concept of a KBCC helps to extend the analysis beyond localized geographies because it conceptualizes innovation processes as going beyond any conflation with either a firm or a place. It emphasizes the multi-scalar linkages and relationships formed by different actors, which are, in turn, grounded in specific institutional geographies. In turn, innovation processes are inherently related to these multi-scalar social interactions, something which is most evident in the difficulties that individual agents (e.g. life science firms) encounter in relation to the creation and application of knowledge, especially when it comes to capturing value from knowledge. This has significant implications for organizational fragmentation and governance, which means that earlier conceptions of commodity chain governance, such as the *producer-driven* and *consumer-driven* models developed by Gereffi (2001), are inadequate for considering the concentration and dispersal of innovation processes necessary to knowledge-based sectors because these governance forms ignore the collaborative inter-dependence of numerous organizations (e.g. universities, small innovative firms, large multi-nationals, etc.). As mentioned above, a new form of *alliance-driven governance* characterizes KBCC (Birch, 2008) – see Table 2.2 for details.

On the organizational and institutional level, alliance-driven governance necessitates core competencies in collaborating with and acquiring and absorbing knowledge from different organizations (Senker, 2005), as well the competency to operate across different, and often distinct, institutional and regulatory regimes in order to access knowledge and markets (Ossenbrugge and Zeller, 2002). As a consequence of the attendant uncertainty of these interactions, and the coordination of diverse incentives, organizations are reliant upon financial investment that is neither short-term nor risk-averse; they therefore depend on public funding and high-risk investors such as venture capitalists (Casper and Kettler, 2001). Uncertainty also results from the high asset specificity of new scientific knowledge whose contribution to value creation is often unclear and intangible (e.g. intellectual), which means that innovation depends on new institutional forms of (intellectual) property protection to encourage investment (Arora and Merges, 2004). Thus organizations are tied into diverse institutional contexts entailing varied and numerous external linkages (e.g. to venture capitalist, to government regulators, etc.) that are, themselves, constituted in relation to other institutional contexts (e.g. with extra-regional and extra-national contexts). What this means is that there is a need to consider the institutional context (and its relation to other contexts) of KBCC in more depth, especially if we want to examine how this relates to the capacity of less-favoured regions to develop new, knowledge-based sectors.

Conclusion

The preceding theoretical discussion is meant to highlight the need to go beyond localized analytical approaches for understanding the geography of innovation, especially when it comes to knowledge-based sectors like the life sciences. As the subsequent chapters illustrate, the KBCC framework helps to open up a number of theoretical and empirical avenues for understanding the life sciences. In Chapter 3, I aim to analyze the knowledge–space dynamic in more detail. This chapter problematizes the idea that knowledge and knowledge spillovers are locally constituted, whether identified as codified or tacit forms.

As a result, this opens up space to examine the life sciences in multi-scalar terms in Chapter 4 through applying the KBCC approach I outlined here. This sort of analysis highlights the range of innovation governance at play in the life sciences, especially differences at different scales. I turn to one important aspect of this governance in Chapter 5 when I consider the role of discourse in *constructing* the life sciences sector, especially at the supranational scale. Discourse plays a key role, for example, in shaping and reshaping institutions such that they come to

Table 2.2 Alliance-driven governance (ADG) model

Model	*Characteristics*	*Theoretical underpinnings*	*Spatial implications*
GCC drivers	'Patient capital'	High asset specificity of new science and technology leads to risk and uncertainty and therefore discourages short-term, low-risk investment necessitating public funding and venture capital (VC).	Embedding of investment in the institutional arrangement of national and regional governments (and their agencies) and private capital (e.g. VC).
Core competencies	Collaborating, regulations	Specialization and complexity preclude integration so organizations rely on collaborations that cross national regulatory regimes necessitating an understanding of different regulatory standards.	Dispersal of the innovation process between and across organizations in different regions and countries.
Entry barriers	Economies of complexity	High-cost, analytical knowledge (i.e. science) infrastructure and diverse national regulatory policies inhibit entry.	Concentration of scientific capacity and linkages within broader global knowledge pipelines.
Sectors	High-technology, intangibles	Sectors dependent upon intellectual property protection to ensure value capture and pursuit of regulatory arbitrage.	Dependence on national (e.g. patent offices) and global (e.g. WTO) governance institutions.
Network linkages	Alliance based	Requirements of collaborating and regulatory adherence mean that organizations rely on the coordination of diverse incentives.	Inter-organizational and cross-border knowledge, resource and regulatory linkages and coordination.
Network structure	Matrix	The collective nature of innovation and high asset specificity mean that networks consist of numerous interactions.	Multi-organizational, multi-scalar and dynamic knowledge networks.

Source: adapted from Birch (2008).

fulfil (or sometimes not) the discursive promises being made. As can be expected, however, this sort of social construction comes up against hard financial reality when sectors like the life sciences face major crises, like the global financial crisis. In Chapter 6 I examine how this crisis has impacted on the life sciences as a result of the increasing financialization of the sector.

References

Acharya, R. (1999) *The Emergence and Growth of Biotechnology*, Cheltenham: Edward Elgar.

Acharya, R., Arundel, A. and Orsenigo, L. (1998) The evolution of European biotechnology and its future competitiveness, in J. Senker (ed.) *Biotechnology and Competitive Advantage*, Cheltenham: Edward Elgar.

Acs, Z., Audretsch, D. and Feldman, M. (1991) Real effects of academic research: comment, *American Economic Review* 82: 363–367.

Amin, A. and Thrift, N. (1992) Neo-Marshallian nodes in global networks, *International Journal of Urban and Regional Research* 16: 571–587.

Amin, A. and Thrift, N. (1994) Living in the global, in A. Amin and N. Thrift (eds) *Globalization, Institutions, and Regional Development in Europe*, Oxford: Oxford University Press.

Arora, A. and Merges, R. (2004) Specialized supply firms, property rights and firm boundaries, *Industrial and Corporate Change* 13(3): 451–475.

Asheim, B. and Coenen, L. (2006) The role of regional innovation systems in a globalising economy, in G. Vertova (ed.) *The Changing Economic Geography of Globalization*, London: Routledge.

Asheim, B. and Gertler, M. (2005) The geography of innovation: Regional innovation systems, in J. Fagerberg, D. Mowery and R. Nelson (eds) *The Oxford Handbook of Innovation*, Oxford: Oxford University Press.

Audretsch, D. and Stephan, P. (1996) Company–scientist locational links: The case of biotechnology, *The American Economic Review* 86: 641–652.

Bagchi-Sen, S., Hall, L. and Petryshyn, L. (2001) A study of university–industry linkages in the biotechnology industry: perspectives from Canada, *International Journal of Biotechnology* 3(3/4): 390–410.

Bair, J. (2005) Global capitalism and commodity chains: Looking back, going forward, *Competition and Change* 9(2): 153–180.

Bathelt, H. and Gluckler, J. (2005) Resources in economic geography: From substantive concepts towards a relational perspective, *Environment and Planning A* 37: 1545–1563.

Bathelt, H., Malmberg, A. and Maskell, P. (2004) Clusters and knowledge: Local buzz, global pipelines and the process of knowledge creation, *Progress in Human Geography* 28: 31–56.

Bell, D. (1973) *The Coming of Post-Industrial Society*, New York: Basic Books.

Benner, M. and Lofgren, H. (2007) The bio-economy and the competition state: Transcending the dichotomy between coordinated and liberal market economies, *New Political Science* 29(1): 77–95.

Birch, K. (2006) The neoliberal underpinnings of the bioeconomy: The ideological discourses and practices of economic competitiveness, *Genomics, Society and Policy* 2(3): 1–15.

Birch, K. (2007) The social construction of the biotech industry, in P. Glasner, P. Atkinson and H. Greenslade (eds) *New Genetics, New Social Formations*, London: Routledge, pp. 94–113.

Birch, K. (2008) Alliance-driven governance: Applying a global commodity chains approach to the UK biotechnology industry, *Economic Geography* 84(1): 83–103.

Birch, K. (2011) 'Weakness' as 'strength' in the Scottish life sciences: Institutional embedding of knowledge-based commodity chains in a less-favoured region, *Growth and Change* 42(1): 71–96.

Birch, K. (2012) Knowledge, place and power: Geographies of value in the bioeconomy, *New Genetics and Society* 31(2): 183–201.

Birch, K. (2015) *We Have Never Been Neoliberal: A Manifesto for a Doomed Youth*, Winchester: Zero Books.

Birch, K. and Cumbers, A. (2010) Knowledge, space and economic governance: The implications of knowledge-based commodity chains for less-favoured regions, *Environment and Planning A* 42(11): 2581–2601.

Birch, K. and Mykhnenko, V. (2014) Lisbonizing vs. financializing Europe? The Lisbon Strategy and the (un-)making of the European knowledge-based economy, *Environment and Planning C* 32(1): 108–128.

Boschma, R. (2005) Proximity and innovation: A critical assessment, *Regional Studies* 39: 61–74.

Braczyk, H.-J., Cooke, P. and Heidenreich, M. (eds) (1998) *Regional Innovation Systems*, London: UCL Press.

Breschi, S., Lissoni, F. and Orsenigo, L. (2001) Success and failure in the development of biotechnology clusters: The case of Lombardy, in G. Fuchs (ed.) *Comparing the Development of Biotechnology Clusters*, London: Harwood Academic Publishers.

Brint, S. (2001) Professionals and the 'Knowledge Economy': Rethinking the theory of Postindustrial Society, *Current Sociology* 49(4): 101–132.

Bunnell, T. and Coe, N. (2001) Spaces and scales of innovation, *Progress in Human Geography* 25(4): 569–589.

Camagni, R. (1995) The concept of *innovative milieu* and its relevance for public policies in European lagging regions, *Papers in Regional Science* 74: 317–340.

Casper, S. and Karamos, A. (2003) Commercializing science in Europe: The Cambridge biotechnology cluster, *European Planning Studies* 11: 805–822.

Casper, S. and Kettler, H. (2001) National institutional frameworks and the hybridization of entrepreneurial business models: The German and UK biotechnology sectors, *Industry and Innovation* 8: 5–30.

Casper, S. and Murray, F. (2004) Examining the marketplace for ideas: How local are Europe's biotechnology clusters?, in M. McKelvey, A. Rickne and J. Laage-Hellman (eds) *The Economic Dynamic of Modern Biotechnology*, Cheltenham: Edward Elgar.

Chakrabarti, A. and Weisenfeld, U. (1991) An empirical analysis of innovation strategies of biotechnology firms in the US, *Journal of Engineering and Technology Management* 8: 243–260.

Coe, N., Hess, M., Yeung, H., Dicken, P. and Henderson, J. (2004) 'Globalizing' regional development: A global production networks perspective, *Transactions of the Institute of British Geographers* NS 29: 468–484.

Coenen, L., Moodysson, J. and Asheim, B. (2004) Nodes, networks and proximities: On the knowledge dynamics of the Medicon Valley biotech cluster, *European Planning Studies* 12: 1003–1018.

Cooke, P. (1998) Introduction: Origins of the concept, in H.-J. Braczyk, P. Cooke and M. Heidenreich (eds) *Regional Innovation Systems*, London: UCL Press.

Cooke, P. (2001) Regional innovation systems, clusters, and the knowledge economy, *Industrial and Corporate Change* 10: 945–974.

Cooke, P. (2002) *Knowledge Economies*, London: Routledge.
Cooke, P. (2003a) The evolution of biotechnology in three continents: Schumpeterian or Penrosian?, *European Planning Studies* 11: 757–763.
Cooke, P. (2003b) Geographic clustering in the UK biotechnology sector, in G. Fuchs (ed.) *Biotechnology in Comparative Perspective*, London: Routledge.
Cooke, P. (2004a) The molecular biology revolution and the rise of bioscience megacentres in North America and Europe, *Environment and Planning C* 22: 161–177.
Cooke, P. (2004b) Life sciences clusters and regional science policy, *Urban Studies* 41: 1133–1131.
Cooke, P. (2004c) Regional knowledge capabilities, embeddedness of firms and industry organisation: Bioscience megacentres and economic geography, *European Planning Studies* 12: 625–641.
Cooke, P. (2004d) Introduction: Regional innovation systems – an evolutionary approach, in P. Cooke, M. Heidenreich and H.-J. Braczyk (eds) *Regional Innovation Systems* (2nd Edition), London: Routledge.
Cooke, P. (2006) Introduction: Regional asymmetries, knowledge categories and innovation intermediation, in P. Cooke and A. Piccaluga (eds) *Regional Development in the Knowledge Economy*, London: Routledge.
Cooke, P. and Leydesdorff, L. (2006) Regional development in the knowledge-based economy: The construction of advantage, *Journal of Technology Transfer* 31: 5–15.
Cooke, P. and Morgan, K. (1998) *The Associational Economy: Firms, Regions, and Innovation*, Oxford: Oxford University Press.
Coombs, R., Saviotti, P. and Walsh, V. (1987) *Economics and Technological Change*, London: Macmillan.
Deeds, D. and Hill, C. (1996) Strategic alliances and the rate of new product development: An empirical study of entrepreneurial biotechnology firms, *Journal of Business Venturing* 11: 41–55.
della Valle, F. and Gambardella, A. (1993) 'Biological' revolution and strategies for innovation in pharmaceutical companies, *R&D Management* 23: 287–302.
Dodgson, M. (1991) Strategic alignment and organizational options in biotechnology firms, *Technology Analysis & Strategic Management* 3: 115–125.
Dosi, G. (1988) Sources, procedures, and microeconomic effects of innovation, *Journal of Economic Literature* 26: 1120–1171.
DTI (1999a) *Biotechnology Clusters Report*, London: Department of Trade and Industry.
DTI (1999b) *Genome Valley: The Economic Potential and Strategic Importance of Biotechnology in the UK*, London: Department of Trade and Industry.
Dutfield, G. (2003) *Intellectual Property Rights and the Life Science Industries*, Hampshire: Ashgate.
Ernst, D. and Kim, L. (2002) Global production networks, knowledge diffusion, and local capability formation, *Research Policy* 31: 1417–1429.
European Council (2000) *An Agenda of Economic and Social Renewal for Europe* (aka Lisbon Agenda), Brussels: European Council [DOC/00/7].
Fagerberg, J. (2005) Innovation: A guide to the literature, in J. Fagerberg, D. Mowery and R. Nelson (eds) *The Oxford Handbook of Innovation*, Oxford: Oxford University Press, pp. 1–26.
Fagerberg, J. and Verspagen, B. (2009) Innovation studies – The emerging structure of a new scientific field, *Research Policy* 38: 218–233.
Fagerberg, J., Martin, B. and Anderson, E. (eds) (2013) *Innovation Studies: Evolution and Future Challenges*, Oxford: Oxford University Press.

Feldman, M. (1999) The new economics of innovation, spillovers and agglomeration: A review of empirical studies, *Economic Innovation and New Technology* 8: 5–25.

Feldman, M. (2000) Location and innovation: The new economic geography of innovation, spillovers, and agglomeration, in G. Clark, M. Feldman and M. Gertler (eds) *The Oxford Handbook of Economic Geography*, Oxford: Oxford University Press.

Feldman, M. and Francis, J. (2003) Fortune favours the prepared region: The case of entrepreneurship and the Capitol region biotechnology cluster, *European Planning Studies* 11: 765–788.

Freeman, C., Clark, J. and Soete, L. (1982) *The Economics of Industrial Innovation*, London: Pinter.

Fuller, S. (2001) A critical guide to knowledge society newspeak: Or, how not to take the great leap backward, *Current Sociology* 49: 177–201.

Gereffi, G. (1994) The organization of buyer-driven global commodity chains: How U.S. retailers shape overseas production networks, in G. Gereffi and M. Korzeniewicz (eds) *Commodity Chains and Global Capitalism*, Westport, CT: Greenwood Press, pp. 95–122.

Gereffi, G. (1996) Global commodity chains: New forms of coordination and control among nations and firms in international industries, *Competition and Change* 1(4): 427–439.

Gereffi, G. (2001) Beyond the producer-driven/buyer-driven dichotomy: The evolution of global value chains in the internet era, *IDS Bulletin* 32(3): 30–40.

Gereffi, G., Humphrey, J. and Sturgeon, T. (2005) The governance of global value chains, *Review of International Political Economy* 12: 78–104.

Gertler, M. and Levitte, Y. (2005) Local nodes in global networks: The geography of knowledge flows in biotechnology innovation, *Industry and Innovation* 12: 487–507.

Godin, B. (2006a) The knowledge-based economy: Conceptual framework or buzzword?, *Journal of Technology Transfer* 31: 17–30.

Godin, B. (2006b) The linear model of innovation: The historical construction of an analytical framework, *Science, Technology and Human Values* 31(6): 639–667.

Godin, B. (2012) 'Innovation Studies': The invention of a specialty, *Minerva* 50: 397–421.

Godin, B. (2014) 'Innovation Studies': Staking the claim for a new disciplinary 'tribe', *Minerva* 52: 489–495.

Godin, B. (2015) *Innovation Contested*, London: Routledge.

Government of Canada (2014) *Seizing Canada's Moment*, Ottawa: Industry Canada.

Granovetter, M. (1985) Economic action and social structure: the problem of embeddedness, *American Journal of Sociology* 91: 481–510.

Gray, M. and Parker, E. (1998) Industrial change and regional development: The case of the US biotechnology and pharmaceutical industries, *Environment and Planning A* 30: 1757–1774.

Greis, N., Dibner, M. and Bean, A. (1995) External partnering as a response to innovation barriers and global competition in biotechnology, *Research Policy* 24: 609–630.

Hall, L. and Bagchi-Sen, S. (2001) A study of R&D, innovation and business performance in the Canadian biotechnology industry, *Technovation* 22: 231–244.

Hall, P. and Soskice, D. (eds) (2001) *Varieties of Capitalism: The Institutional Foundations of Comparative Advantage*, Oxford: Oxford University Press.

Hamilton, W., Villa, J. and Dibner, M. (1990) Patterns of choice in emerging firms: Positioning for innovation in biotechnology, *California Management Review* 32: 73–86.

Harvey, D. (1999[1982]) *The Limits to Capital*, London: Verso.

Henderson, J., Dicken, P., Hess, M., Coe, N. and Yeung, H. (2002) Global production networks and the analysis of economic development, *Review of International Political Economy* 9: 436–464.

Henry, N. and Pinch, S. (2000) (The) industrial agglomeration (of Motor Sport Valley): A knowledge, space, economy approach, in J. Bryson, P. Daniels, N. Henry and J. Pollard (eds) *Knowledge, Space, Economy*, London: Routledge.

Hess, M. and Yeung, H. (2006) Guest editorial: Whither global production networks in economic geography?, *Environment and Planning A* 38: 1193–1204.

HM Treasury (2014) *Our Plan for Growth: Science and Innovation*, London: HM Treasury and Department for Business, Innovation and Skills.

Hudson, R. (2005) Rethinking change in old industrial regions: Reflecting on the experiences of North East England, *Environment and Planning A* 37: 581–596.

Jessop, B. (2000) The state and the contradictions of the knowledge-driven economy, in J. Bryson, P. Daniels, N. Henry and J. Pollard (eds) *Knowledge, Space, Economy*, London: Routledge.

Kettler, H. and Casper, S. (2000) *The Road to Sustainability in the UK and German Biotechnology Industries*, London: Office of Health Economics.

Kline, S. and Rosenberg, N. (1986) An overview of innovation, in R. Landau and N. Rosenberg (eds) *The Positive Sum Strategy*, Washington DC: National Academy Press.

Lagendijk, A. (2006) Learning from conceptual flow in regional studies: Framing present debates, unbracketing past debates, *Regional Studies* 40: 385–399.

Lawton-Smith, H., Mihell, D. and Kingham, D. (2000) Knowledge-complexes and the locus of technological change: The biotechnology sector in Oxfordshire, *Area* 32: 179–188.

Leibovitz, J. (2004) 'Embryonic' knowledge-based clusters and cities: The case of biotechnology in Scotland, *Urban Studies* 41: 1133–1155.

Loeppky, R. (2004) International restructuring, health and the advanced industrial state, *New Political Economy* 9: 493–513.

Loeppky, R. (2005) History, technology, and the capitalist state: The comparative political economy of biotechnology and genomics, *Review of International Political Economy* 12: 264–286.

Lovering, J. (1999) Theory led by policy: The inadequacies of 'The New Regionalism', *International Journal of Urban and Regional Research* 23: 379–395.

Lundvall, B.-A. (ed.) (1992) *National Systems of Innovation*, London: Pinter.

Machlup, F. (1962) *The Production and Distribution of Knowledge in the US*, Princeton: Princeton University Press.

MacKinnon, D., Cumbers, A. and Chapman, K. (2002) Learning, innovation and regional development: a critical appraisal of recent debates, *Progress in Human Geography* 26: 293–311.

MacLeod, G. (2001) New regionalism reconsidered: Globalization and the remaking of political economic space, *International Journal of Urban and Regional Research* 25: 804–829.

Malecki, E. (2000) Creating and sustaining competitiveness: Local knowledge and economic geography, in J. Bryson, P. Daniels, N. Henry and J. Pollard (eds) *Knowledge, Space, Economy*, London: Routledge.

Malmberg, A. (2003) Beyond the cluster – Local milieus and global connections, in J. Peck and H. Yeung (eds) *Remaking the Global Economy*, London: SAGE.

Malmberg, A. and Maskell, P. (2002) The elusive concept of localization economies: Towards a knowledge-based theory of spatial clustering, *Environment and Planning A* 34: 429–449.

Malmberg, A. and Power, D. (2005) (How) do (firms in) clusters create knowledge?, *Industry and Innovation* 12: 409–431.

Martin, R. and Sunley, P. (2003) Deconstructing clusters: Chaotic concept or policy panacea?, *Journal of Economic Geography* 3: 5–35.

Massey, D. (1995[1984]) *Spatial Divisions of Labour*, London: Macmillan.

May, C. (2000) *A Global Political Economy of Intellectual Property Rights*, London: Routledge.

Mirowski, P. and Sent, E.-M. (2008) The commercialization of science and the response of STS, in E. Hackett, O. Amsterdamska, M. Lynch and J. Wajcman (eds) *Handbook of Science and Technology Studies*, Cambridge MA: MIT Press, pp. 635–689.

Moodysson, J., Coenen, L. and Asheim, B. (2008) Explaining spatial patterns of innovation: Analytical and synthetic modes of knowledge creation in the Medicon Valley life-science cluster, *Environment and Planning A* 40(5): 1040–1056.

Morgan, K. (1997) The learning region: Institutions, innovation and regional renewal, *Regional Studies* 31: 491–503.

Moulaert, F. and Sekia, F. (2003) Territorial innovation models: A critical survey, *Regional Studies* 37: 289–302.

Nelson, R. and Winter, S. (1982) *An Evolutionary Theory of Economic Change*, London: Belknap Harvard.

Nightingale, P. and Martin, P. (2004) The myth of the biotech revolution, *Trends in Biotechnology* 22: 564–569.

Niosi, J. and Bas, T. (2004) Canadian biotechnology policy: Designing incentives for a new technology, *Environment and Planning C* 22: 233–248.

OECD (1996) *The Knowledge-Based Economy*, Paris: Organisation for Economic Co-operation and Development.

Ossenbrugge, J. and Zeller, C. (2002) The biotech region of Munich and the spatial organisation of its innovation networks, in L. Schätzl and J. Revilla (eds) *Technological Change and Regional Development in Europe*, Berlin: Physica-Verlag.

Owen-Smith, J. and Powell, W. (2003) The expanding role of university patenting in the life sciences: Assessing the importance of experience and connectivity, *Research Policy* 32(9): 1695–1711.

Owen-Smith, J., Riccaboni, M., Pamolli, F. and Powell, W. (2002) A comparison of US and European university–industry relations in the life sciences, *Management Science* 48: 24–43.

Pike, A., Rodríguez-Pose, A. and Tomaney, J. (2006) *Local and Regional Development*, London: Routledge.

Pinch, S., Henry, N., Jenkins, M. and Tallman, S. (2003) From 'industrial districts' to 'knowledge clusters': A model of knowledge dissemination and competitive advantage in industrial agglomerations, *Journal of Economic Geography* 3: 373–388.

Pisano, G. (2006) *Science Business*, Cambridge, MA: Harvard University Press.

Porter, M. (1990) *The Competitive Advantage of Nations*, London: Macmillan.

Porter, M. (2000) Location, competition, and economic development: Local clusters in a global economy, *Economic Development Quarterly* 14: 15–34.

Powell, W. (1998) Learning from collaboration: knowledge and networks in the biotechnology and pharmaceutical industries, *California Management Review* 40: 228–240.

Powell, W. and Snellman, K. (2004) The knowledge economy, *Annual Review of Sociology* 30: 199–220.

Powell, W., Koput, K. and Smith-Doerr, L. (1996) Interorganizational collaboration and the locus of innovation: Networks of learning in biotechnology, *Administrative Science Quarterly* 41: 116–145.

Powell, W., Koput, K., Bowie, J. and Smith-Doerr, L. (2002) The spatial clustering of science and capital: Accounting for biotech firm–venture capital relationships, *Regional Studies* 36: 291–305.

Powell, W., White, D., Koput, K. and Owen-Smith, J. (2004) Network dynamics and field evolution: The growth of interorganizational collaboration in the life sciences, *American Journal of Sociology* 110(4): 1132–1205.

Prevezer, M. (1997) The dynamics of industrial clustering in biotechnology, *Small Business Economics* 9: 255–271.

Prevezer, M. and Toker, S. (1996) The degree of integration in strategic alliances in biotechnology, *Technology Analysis & Strategic Management* 8: 117–133.

Rosenberg, N. (1976) *Perspectives on Technology*, Cambridge: Cambridge University Press.

Rosiello, A. (2004) *Evaluating Scottish Enterprise's Cluster Policy in Life Sciences: A Descriptive Analysis*, University of Edinburgh: Innogen Working Paper No.16.

Ryan, C. and Phillips, P. (2004) Knowledge management in advanced technology industries: An examination of international agricultural biotechnology clusters, *Environment and Planning C* 22: 217–232.

Saviotti, P. (1998) Industrial structure and the dynamics of knowledge generation in biotechnology, in J. Senker (ed.) *Biotechnology and Competitive Advantage*, Cheltenham: Edward Elgar.

Saviotti, P., Joly, P.-B., Estades, J., Ramani, S. and de Looze, M.-A. (1998) The creation of European dedicated biotechnology firms, in J. Senker (ed.) *Biotechnology and Competitive Advantage*, Cheltenham: Edward Elgar.

Schumpeter, J. (1939) *Business Cycles: Volume 1*, London: McGraw-Hill Book Company.

Schumpeter, J. (1942) *Capitalism, Socialism, and Democracy* (3rd edition), New York: Harper & Row.

Scott, A. (1989) High technology industry and territorial development: The rise of the Orange County complex, 1955–1984, *Urban Geography* 7: 3–45.

Scott, A. (1998) The geographic foundations of performance, in A. Chandler Jr., P. Hagstrom and O. Solvell (eds) *The Dynamic Firm: The Role of Technology, Strategy, Organization, and Regions*, Oxford: Oxford University Press.

Senker, J. (ed.) (1998) *Biotechnology and Competitive Advantage*, Cheltenham: Edward Elgar.

Senker, J. (2004) An overview of biotechnology innovations in Europe: Firms, demand, government policy and research, in M. McKelvey, A. Rickne and J. Laage-Hellman (eds) *The Economic Dynamics of Modern Biotechnology*, Cheltenham: Edward Elgar.

Senker, J. (2005) *Biotechnology Alliances in the European Pharmaceutical Industry: Past, Present and Future*, SPRU, University of Sussex: SEWPS Paper No. 137.

Senker, J. and Faulkner, W. (1996) Networks, tacit knowledge and innovation, in R. Coombs, A. Richards, P. Saviotti and V. Walsh (eds) *Technological Collaboration*, Cheltenham: Edward Elgar.

Sharp, M. and Senker, J. (1999) European biotechnology: Learning and catching-up, in A. Gambardella and F. Malerba (eds) *The Organization of Economic Innovation in Europe*, Cambridge: Cambridge University Press.

Simmie, J. (2003) Innovation and urban regions as national and international nodes for the transfer and sharing of knowledge, *Regional Studies* 37: 607–620.

Simmie, J. (2004) Innovation and clustering in the globalised international economy, *Urban Studies* 41: 1095–1112.

Smith, N. (2008[1984]) *Uneven Development*, Athens: University of Georgia Press.

Sokol, M. (2004) The 'knowledge economy': A critical view, in P. Cooke and A. Piccaluga (eds) *Regional Economies as Knowledge Laboratories*, Cheltenham: Edward Elgar.

Solow, R. (1956) A contribution to the theory of economic growth, *Quarterly Journal of Economics* 70: 65–94.

Storper, M. (1995) The resurgence of regional economies, ten years later: The region as a nexus of untraded interdependencies, *European Urban and Regional Studies* 2: 191–221.

Storper, M. and Scott, A. (1995) The wealth of regions: Market forces and policy imperatives in local and global context, *Futures* 27: 505–526.

Storper, M. and Walker, R. (1989) *The Capitalist Imperative*, Oxford: Blackwell.

Thompson, P. (2004) *Skating on Thin Ice: The knowledge economy myth*, Glasgow: Big Thinking.

von Hippel, E. (1988) *The Sources of Innovation*, Oxford: Oxford University Press.

von Hippel, E. (1994) 'Sticky information' and the locus of problem solving: Implications for innovation, *Management Science* 40: 429–439.

Walcott, S. (2001) Growing global: Learning locations in the life sciences, *Growth and Change* 32: 511–532.

Walsh, V., Niosi, J. and Mustar, P. (1995) Small-firm formation in biotechnology: A comparison of France, Britain and Canada, *Technovation* 15: 303–327.

Woiceshyn, J. (1995) Lessons in product innovation: A case study of biotechnology firms, *R&D Management* 25: 395–409.

Yoxen, E. (1981) Life as productive force: Capitalising the science and technology of molecular biology, in L. Levidow and B. Young (eds) *Science, Technology and the Labour Process*, London: CSE Books.

Zeller, C. (2001) Clustering biotech: A recipe for success? Spatial patterns of growth of biotechnology in Munich, Rhineland and Hamburg, *Small Business Economics* 17: 123–141.

Zeller, C. (2004) North Atlantic innovative relations of Swiss pharmaceuticals and the proximities with regional biotech arenas, *Economic Geography* 80: 83–111.

Zucker, L., Darby, M. and Armstrong, J. (2002) Commercializing knowledge: University science, knowledge capture, and firm performance in biotechnology, *Management Science* 48: 138–153.

Zucker, L., Darby, M. and Brewer, M. (1998) Intellectual human capital and the birth of US biotechnology enterprises, *The American Economic Review* 88: 291–306.

3 Innovation geographies in the UK life sciences

Introduction

Taking the life sciences seriously as a sector raises the question of why it is spread so unevenly. And in my case, this is as relevant for the UK as anywhere. A considerable amount of academic research has sought to explain this uneven development and, more specifically, its concentration in specific places (for reviews see Feldman 1999; Senker 2005; Birch 2007a). As discussed in the last chapter, the theories can be crudely split in a number of different research agendas. For example, early research on the life sciences drew on strategic management theories to explore the importance of collaborations in the life sciences sector and can be seen as *firm-centred* approaches. The more *system-centred* research in the field of innovation studies and political economy focused on the importance of sectoral and institutional conditions to the life sciences. A final set of *cluster-centred* approaches – perhaps the most influential – drew on the work of Michael Porter (1990) and Bracyzk *et al.* (1998), although the latter's distinct and more sophisticated regional innovation system approach also incorporates a *systems* perspective. Despite the respective insights of these different research agendas, there are a number of gaps in their analysis. More recent approaches in economic geography, for example, have highlighted the importance of *local–global* linkages for life sciences firms (e.g. Gertler and Levitte 2005; Birch 2008; Birch and Cumbers 2010), which I will come back to in the following chapter.

As I pointed out at the end of the last chapter, my objective here is to analyze the *knowledge–space dynamic* in the life sciences sector, focusing empirically on the UK. I do so in order to engage critically with the various theoretical arguments about the role of knowledge and innovation in regional economic development and growth. I am especially interested in understanding the differences between codified and tacit knowledge forms – identified by Michael Polanyi (1966) – in regional development, as this was a key debate at the time when I undertook this research. Scholars like Markusen (1996), Malecki (1997), Howells (2000, 2002), Malmberg and Maskell (2002) and Gertler (2003), for example, argued that codified knowledge (e.g. patents) is supposedly 'ubiquitous' and therefore easy to transfer and/or acquire, while they also stressed that tacit knowledge is embodied and embedded in local institutions, practices, relations, learning, etc. As these scholars pointed out, Polanyi originally conceptualized the relation between codified and tacit knowledge as a continuum rather than dichotomy, and that the use of codified knowledge depends on tacit knowledge. Consequently, knowledge could actually be *and* is highly geographically concentrated and unevenly distributed. This *knowledge–space dynamic*, as I conceptualize it, helps to explain the emergence of the life sciences in particular places around the UK.

In the rest of this chapter, I bring together the sometimes disparate approaches used to understand the *knowledge–space dynamic* in the life sciences (Birch 2007b). My objective is to incorporate insights from each strand of research not only to illustrate the importance of the geographies of knowledge and innovation, but also to understand how geographical specificity has impacted on the development of the UK life sciences sector. This chapter first reviews the literature on knowledge and spatial dynamics in biotechnology innovation. Next, it uses primary and secondary data on the UK biotech industry to illustrate the extent and geographical patterns of these knowledge and spatial dynamics in order to analyze the relationship between them. This analysis illustrates the geographical differences between the regional concentrations of the UK life sciences. Finally, the chapter concludes by discussing the importance of these patterns in the *knowledge–space dynamic* and the impacts these processes have on the regional development of the life sciences.

Innovation and the *knowledge–space dynamic*

As mentioned in the introduction and the previous chapter, a range of theories and concepts have been used to analyze the relationship between geography and innovation in the life sciences. Here I discuss three major research agendas: firm-centred, system-centred and cluster-centred theories. I then bring them together with my notion of the knowledge–space dynamic, which I subsequently use to analyze the UK life sciences.

Knowledge processes in life sciences innovation

First off, *firm-centred* approaches originate in strategic management and economic sociology concentrating, in particular, on the importance of formal relationships between life sciences firms and other organizations, especially in terms of the impact of alliances (e.g. Deeds and Hill 1996), collaborations (e.g. Chakrabarti and Weisenfeld 1991) and networks (e.g. Powell *et al.* 1996). The underlying concern of such research is how these relationships affect the organization of life sciences research in a firm's innovation strategies and what the complementarities in such linkages mean for both emergent life sciences firms and more established large pharmaceutical companies (see Feldman 1999). This integration of disparate capabilities has led to an interest in inter-firm networks, exemplified in the work of Powell and colleagues (e.g. Powell *et al.* 1996; Powell 1998; Powell *et al.* 2002), where the 'locus of innovation' is seen as the nexus of collaboration itself. In terms of the *knowledge–space dynamic*, such collaborations constitute the innovation process in the individual firm or organization providing the means for the accretion of organizational capabilities through the absorption of new knowledge and its integration with existing, internal expertise.

The absorption of new knowledge is facilitated by shared rules and routines, meaning that the institutional environment of a firm or organization is crucial as research in *systems-centred* approaches highlights. Here the emphasis on evolutionary and Schumpeterian theories of innovation, alongside historical theories of technological change, places an emphasis on the systemic nature of innovation concerned with national institutional characteristics and firm-level competitiveness (e.g. Bartholomew 1996; Senker *et al.* 1996; Acharya *et al.* 1998; Saviotti *et al.* 1998). Some authors (e.g. Coriat *et al.* 2003) have argued that the life sciences sector represents a 'new science-based innovation regime'; others have suggested that there is a *Triple Helix* model of innovation connecting academia, industry and government (Etzkowitz and Leydesdorff 2000). The *knowledge–space dynamic* in this context entails the consideration of complementary capabilities to innovation; this is not simply limited to the distinction between 'analytical' or 'synthetic' knowledge inputs (see Moodysson *et al.* 2008), but also commercial knowledge of financing, regulations, market opportunities etc. (see Cooke 2007, 26–31; Kasabov and Delbridge 2008). Furthermore, the contribution of specific complementarities to the innovation process is not only dependent upon the institutional environment of the 'innovator' organization, but also that of the collaborating actor.

Spatial processes in life sciences innovation

Whilst the *firm* and *system* perspectives provide a number of insights into life sciences innovation, they do not provide an adequate explanation for why the life sciences sector is concentrated in particular places. These spatial dynamics are the focus of *cluster-centred* research on life sciences innovation, which was stimulated by 'new economic geography' (NEG) and work on the concept of knowledge spillovers as an explanation for agglomeration economies (e.g. Acs *et al.* 1991; Audretsch and Feldman 1996). The resurgent interest in *space* – as opposed to *place* – in NEG highlighted the relationship between the location of academic scientists and life sciences firms (Audretsch and Stephan 1996, 1999; Zucker *et al.* 1998; Zucker *et al.* 2002). This perspective grounds the *knowledge–space dynamic* in the specific location of innovating firms and other organizations because it is their co-location with one another that provides access to complementary capabilities, not only in terms of formal collaborations, but also knowledge 'leakage' between organizations.

Similarly, although distinct in terms of the explanation for *clustering*, the work of Porter (1990) on clusters and Braczyk *et al.* (1998) on regional innovation systems (RIS) has encouraged interest in the geographical concentration of life sciences innovation. Both cluster theory, which has proved particularly popular in policy circles (e.g. DTI 1999a), and the distinct RIS perspective have encouraged a broad academic debate in regional studies and economic geography (see Cooke 2001, 2003, 2004a, 2004b, 2005, 2006, 2007; Zeller 2001; Kaiser 2003;

Prevezer 2003; Asheim and Coenen 2006; Birch 2008; Moodysson *et al.* 2008). The more sophisticated RIS approach emphasizes the importance of systems of place-specific organizations and institutions such as firms, business services, public research, education, labour and so on. Furthermore, tacit knowledge and social capital (e.g. trust) embed this array of actors in particular places (Breschi *et al.* 2001; Ryan and Phillips 2004; Cooke 2007) and are, in turn, promoted by such embedding. What this RIS research contributes to understanding the *knowledge–space dynamic* is the idea of iterative learning; that is, through inter-organizational knowledge exchange firms and organizations help to embed place-specific institutional infrastructures which, in turn, encourage further knowledge exchange.

Whilst the *cluster-centred* approaches provide an insight into locally-embedded innovation dynamics, they do not explicitly examine the role of *local–global* linkages (see Bathelt *et al.* 2004). Such extra-local connections and interaction are crucial to life sciences innovation providing diverse knowledge inputs into research and commercialization (Coenen *et al.* 2004; Leibovitz 2004; Ryan and Phillips 2004; Zeller 2004; Gertler and Levitte 2005; Bagchi-Sen 2007; Birch 2008). Thus extra-local connections introduce new knowledge into life sciences innovation and represent a central part of the *knowledge–space dynamic* as they help to explain how the geographical specificity of life sciences innovation depends upon both similarity – in terms of shared institutional environment – and diversity – in terms of extra-local linkages.

The knowledge–space dynamic

As the above discussion of the existing literature on life sciences innovation shows, there is an important dynamic between knowledge and space that is often obscured by concentrating on one or the other. Consequently it is useful to synthesize these debates to understand the *knowledge–space dynamic*.

This dynamic can be summarized in a number of propositions as follows. First, economic performance and development are underpinned by Schumpeterian innovation, although it is contestable whether technological change *per se* is necessarily 'progressive'. Rather it is better to consider innovation as path dependent and therefore constraining the choices available to actors. Second, knowledge provides a crucial input to innovation in that it enables actors to understand the world and make decisions that affect the world. Such knowledge can be acquired from diverse locations and comes in many forms, as well as entailing values, meanings and expectations (i.e. discourses) – which I return to later in the book in Chapter 5 – about what constitutes knowledge and its impact on technological change, economic development, and civil society.

Third, every spatial context is unique meaning that all knowledge entails a geographical specificity in terms of its positioning and embedding in certain places. These can be considered as the knowledge capabilities of different places. Fourth, although knowledge is geographically embedded, it is dependent upon interactions and relationships that cut across multiple scales that further position and embed knowledge in particular places. Fifth, the iterative process of learning resulting from such interactions and relationships is systemic and dependent upon processes that are bounded by and embedded in place-specific organizations, institutions and cultural rules that both enable and constrain people in working across different scales.

What this means is that life sciences innovation is not only shaped by particular scientific and technological regimes and trajectories, but also by the relationship between knowledge and space. Thus the focus on one or the other misses the existing dynamic between them, especially how certain capabilities are embedded in specific geographies that are positioned in relation to other places. Thus it is important to understand how different places of the life sciences differ and what this might mean for regional development.

Political-economic context of the UK life sciences

Before turning to the specific life sciences 'clusters' in the UK, it is worthwhile providing some political-economic context for the analysis. By this I mean a history of the origins of the UK life sciences and some key institutional factors that influence life sciences innovation. For more on the history of the UK life sciences I would recommend reading a couple of new books: *The New Health Bioeconomy* by James Mittra (2016) and *Science, the State and the City* by Geoffrey Owen and Michael Hopkins (2016). They provide more depth than I can do in this chapter. More generally, there is a range of research on the changing political-economic context of the life sciences in the USA and globally (e.g. Birch 2007b; Mirowski 2011).

The origins of the UK life sciences can be traced back to 1978 when the then Labour government commissioned an inquiry into 'biotechnology' in response to fears about the USA steaming ahead in this area (Gottweis 1998a, 1998b). Alfred Spinks, ex-research director at Imperial Chemical Industries (ICI), was asked to head the commission. The inquiry eventually led to the production of a report in 1980, which subsequently became known as the *Spinks Report* (ACARD *et al.* 1980). The commission had the remit to assess the 'existing and prospective science and technology relevant to industrial opportunities in biotechnology' and comprised a working group from the Advisory Council for Applied Research and Development (ACARD), Advisory Board for the Research Councils (ABRC) and the Royal Society (ibid.: foreword). According to Herbert Gottweis (1998a: 196), the composition of the 'Spinks Working Party' indicated a 'new orientation for the political coding of the new biology' towards a concern with the commercial application and benefits of biotechnological sciences as opposed to concerns with health and safety. A dominant theme throughout the report – and one that rears its head again and again over the following decades (e.g. House of Lords 1993; BIGT 2003), as evidenced in Chapter 5 – is the fear of lost market opportunities and national advantage.

After the publication of the *Spinks Report* in 1980, the new Conservative government – elected with a mandate to reduce the role of government in industry – produced a 1981 White Paper as a response. This White Paper broadly claimed that biotechnology could be left to the market and did not require government support (Sharp 1985). Despite these claims, however, several important public initiatives were undertaken between 1980 and 1982 to stimulate biotechnology research and commercialization. First, in 1980 the Conservative government founded a new firm called Celltech, which was given special rights to commercialize research funded by the Medical Research Council (MRC). Second, in 1981 the Science and Engineering Research Council (SERC) established a Biotechnology Directorate to support applied research. Finally, in 1982 the Department of Trade and Industry (DTI) set up a Biotechnology Unit to support commercialization of research (Sharp 1985; Bud 1993).

Despite these specific changes to life sciences research and innovation in the UK, Gottweis (1998a: 203, 205) argued that the Conservative government 'failed' to properly follow up on recommendations from the *Spinks Report*, and took 10 years to 'find a suitable organizational structure for funding decisions in biotechnology'. In contrast to Gottweis, Margaret Sharp (1985) argued that the Conservative government strongly promoted biotechnology across a range of schemes, including funding basic and applied science and encouraging public–private partnerships. However, there was a shift away from central government direction of industry with the reorganization of the National Research Development Corporation (NRDC) and National Enterprise Board (NEB). The NRDC had a monopoly over the commercial exploitation of government-funded research. It was merged with NEB in 1981 to form the British Technology Group (BTG), which then lost its monopoly in 1985 (Owen 2001). BTG was eventually privatized in 1988, and universities were given the right to exploit their research. In 1994, the government established the Biotechnology and Biological Sciences Research Council (BBSRC).

While the public sector was reorganizing the funding and support for biotechnology, the private sector began to take an active role in supporting the establishment of new life sciences firms (see Owen and Hopkins 2016). Examples include:

- The founding of British Biotech in 1986 by scientists from a subsidiary of the US company Searle (since merged into Vernalis in 2003);
- The founding of Enzymatix in 1987 by Christopher Evans, subsequently a serial entrepreneur who had returned from working at Genzyme in the USA (eventually split up into different firms and sold off);
- The founding of several spin-outs by Evans like Celsis International and Chiroscience (the latter merged with Celltech in 1999, which was then taken over by UCB in 2004).

Major changes in the private sector included the 1993 decision by the London Stock Exchange (LSE) to change their listing requirements so that firms could raise public investment without requiring historical revenues, profitability, and trading experience (Gottweis 1998a). Despite introducing other requirements, this change – driven by what was happening in the USA – was meant to enable UK life sciences firms to access long-term, ongoing investment through capital markets. As a result of such changes, by 1995 around 25 life sciences firms had gone public (Owen 2001).

Following the 1997 election of the 'New' Labour government, there was a concerted effort to support high technology sectors like the life sciences. First, the DTI introduced several schemes (e.g. Biotechnology Exploitation Platforms, Biotechnology Mentoring Incubator, etc.) over the following years designed to support the life sciences. Second, the Labour government also changed local and regional planning policy. They introduced new Regional Development Agencies (RDAs) in 1998, which were tasked with developing regional-specific support for new sectors. They also changed *Planning Policy Guidance* (PPG) in 1999, requiring that planning should provide support for cluster development at the regional level (ODPM 2004). Finally, they sought to incentivize: (a) capital investment in life sciences firms through the introduction of Regional Venture Capital (RVC) Funds in 1999; (b) share options for employees through the Enterprise Management Incentive Scheme (EMIS) in 2000; and (c) research and development spending through R&D tax credits in 2000.

Overall, by the early 2000s a number of changes to public and private institutions had been made to support the establishment and growth of the life sciences sector. Not all were necessarily successful, nor without any perverse consequences, but they indicate the extent to which the life sciences has been a matter of concern for a number of governments over the years. It is, in light of these changes, worth considering which regions in the UK have benefited the most from this public (and private) largesse; was it all, or only some parts of the country?

Economic geography of the UK life sciences

Methodological note

As the theoretical discussion above highlighted, there are a number of particular trends that need to be considered in order to understand innovation in the UK life sciences. These can be simply summarized as follows: (1) the extent and size of biotech firms to denote the innovative capacities in a region; (2) the extent and 'strength' of the public science base to indicate regional institutional ecosystem; (3) the extent of knowledge assets (e.g. patents and publications) in a region to illustrate possible spillovers; and (4) the extent of formal alliances and multi-scalar dimensions of knowledge exchange to demonstrate the important role of local–global linkages as a source of diversity. In order to analyze these trends I draw on secondary data collected in 2003–2004 from a number of public and private data sources. These data sources include websites of regional life sciences networks such as *Oxfordshire Bioscience Network*, *BioDundee*, *Eastern Region Biotech Initiative* etc., as well as databases specific to the life sciences produced by *Nature Biotechnology*, *BioWorld* and *BioCommerce*. I also drew on more general databases like *FAME* where information was not available from the other ones. I collected other secondary data from the higher education funding councils and research councils for information on the public science base; patent offices and the *Web of Knowledge* website for information on knowledge assets;

and the *Bioworld* website for information on company alliances (see Birch 2006 for more details).

In order to analyze the knowledge–space dynamic in the UK life sciences, I undertook a series of 108 structured interviews during 2003–2004 to collect data on the dimensions of knowledge exchange, especially as they relate to codified, tacit and commercial knowledge. I use this data here to discuss the specific knowledge and innovation processes of different UK regions, as well as between supposedly 'clustered' life sciences firms and 'non-clustered' ones. The interview sample covered two main groups implicated in the innovation ecosystem: first, 'innovators', defined as people who had directly contributed to the development of a life sciences product/process; and second, 'service providers', defined as people who provide a service to life sciences firms (e.g. management advice, technology parks, finance etc.). The objective of approaching both innovators and service providers was to adopt a systems approach in understanding biotech innovation, which necessitated understanding the actions of diverse social actors (again, see Birch 2006 for more details).

The four centres of the UK life sciences sector

Before unpacking the *knowledge–space dynamic* in the UK life sciences, it is important to outline the uneven regional spread of the sector in the early 2000s. This helps to both illustrate the specificity of regional concentrations in terms of particular trends and patterns, and to show how each regional concentration can be seen as following particular trajectories contingent upon the specifics of the *knowledge–space dynamic* I explore below. I mapped out the concentration of the UK life sciences using the EuroStat's NUTS2 classification to identify specific regional economies. As evident in Figure 3.1, the life sciences sector was concentrated in four main UK regions: East Anglia; Berkshire, Buckinghamshire and Oxfordshire (henceforth Berks *et al.*); Inner London; and Eastern Scotland.[1] There was a particular concentration of the life sciences in the South and East of England, or what has been called the 'golden triangle' of the public science base.

As the theoretical discussion in the previous chapter highlights, the often stylized representation of the life sciences sector as local or regional 'clusters' in the academic and policy debates belies the fact that each region is distinct from the other. These differences result from the different spatial and knowledge trends and processes at play in each region, to which I return below; that is, the regional knowledge–space dynamic helps to explain the uneven spread of the life sciences across the UK. Some regions appeared to have similarities with one another, like East Anglia and Berks *et al.*, especially when contrasted with the 'capital' regions of London and Eastern Scotland. The probable cause of these differences was the relative age of each concentration, where the former have a longer pedigree and history as centres of (life) scientific research – because of their important universities – whereas the latter were (and still are) centres of finance and banking. Despite these superficial similarities, however, each region was distinct from one another as outlined in Table 3.1.

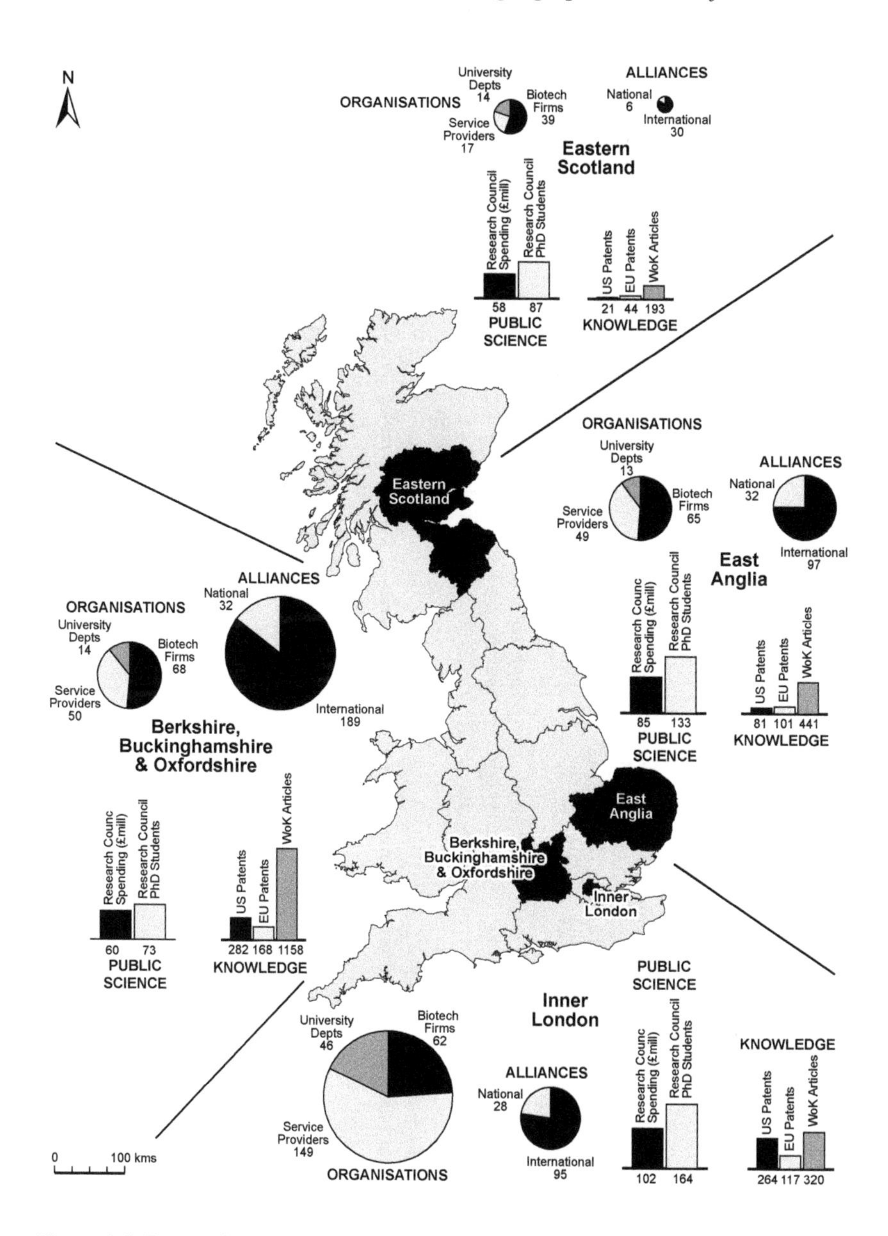

Figure 3.1 Four main concentrations of the UK life sciences sector

Source: produced by Mike Shand and published in Birch (2009); reproduced with permission of Blackwell.

Note: 'WoK' means Web of Knowledge.

Being centred on Cambridge and the university there, East Anglia had a significantly stronger public science base in relation to 'star scientists' and public expenditure than elsewhere. This would imply that East Anglia benefits from

Table 3.1 Knowledge–space characteristics of the main UK life sciences concentrations

REGION	*SPACE*				*KNOWLEDGE*			
	Firms and service	*Public science*	*Scope and scale*	*Origins*	*Public funding*	*Private stocks*	*Labour Market*	*Alliances*
Berks *et al.*	New firms; new services	Old basic; new applied	Micro, medium and large	High local; medium foreign	Medium absolute and relative	High non-appropriable; very high appropriable	High science; very high technical	High total, local and international
East Anglia	Old and new firms; new services	Old and star basic; new applied. Star	Small and medium	Medium local; high foreign	High absolute and relative	Medium non-appropriable; high appropriable	High science and technical	High total and local
Inner London	New firms; old and new services	Old, new and star basic; old applied.	Micro and medium	High local; low foreign	High absolute; low relative	High non-appropriable; medium appropriable	Very high science; medium technical	High total and local
Eastern Scotland	New firms; low services	Old basic and applied	Micro	Low local and foreign	Medium absolute and relative	Low non-appropriable; medium appropriable	High science; low technical	Low

Source: Birch (2009), reproduced with permission of Wiley Blackwell.

knowledge spillovers, as argued in the NEG literature (e.g. Audretsch and Stephan 1996, 1999; Zucker *et al.* 1998; Zucker *et al.* 2002). However, Berks *et al.* – which was a region centred on an equally respected centre of scientific research – had a stronger private science base, represented by higher levels of both appropriable (i.e. patents) and non-appropriable (i.e. publications) knowledge. This would suggest that East Anglia was a more university-centred concentration, while Berks *et al.* was more market-centred. One probable reason for this difference is that there were – at the time of the research – more large firms in Berks *et al.* than East Anglia; according to a number of authors (e.g. Bathelt *et al.* 2004), these organizations provide important global pipelines to other parts of the world.

As is evident from this brief discussion of the four concentrations, they can be characterized in terms of their distinct geographies, which set them apart from one another (and other regions around the world). The specificities of each regional concentration need to be integrated into the analysis of the geography of innovation, implying that generic recommendations built on the findings from one region or another cannot form the basis of some generic regional development policy applicable anywhere in the world. Each place is different, with a different set of political-economic conditions, processes and trends at play.

Knowledge and spatial trends in different life science concentrations

In order to understand the differences between different concentrations in the UK life sciences, it is important to consider different knowledge and spatial patterns and trends. First then, life sciences firms represent the material site of innovation in that they are where knowledge is transformed 'into wealth-creating technologies, products and services through processes of learning and searching' (Asheim and Coenen 2006: 149). As shown above in Figure 3.1, there were four regional concentrations of life sciences firms in the UK in the early 2000s. Each of these regions had over three times the UK average number of life sciences firms ($n = 12$) at the NUTS2 scale and all four represented over half of the UK's life sciences firms ($n = 436$). Most life sciences firms were concentrated in the South and East of England, including London, with two regions alone (East Anglia and Berks *et al.*) containing around a third of all such firms. There were three other NUTS2 regions with above average number of life sciences firms, but they had only around 20 firms each and were therefore not treated as significant concentrations.

It is also useful to consider the concentration of larger life sciences firms since these firms are more likely to play a regional 'anchoring' role (Feldman 2003). Nearly half of the 33 identifiable large firms (over 500 employees) were based in just two regions: East Anglia and Berks *et al.* Interestingly, only 42 per cent of the 107 identifiable medium and large biotech firms (over 250 employees) were based in the four main concentrations, suggesting that these were less dependent on local agglomeration economies. This difference could have been the result of first-mover advantage in East Anglia and Berks *et al.* in that the firms in these two regions were older and therefore more likely to be larger. Thus these two regions could have benefited from the reduction of knowledge fragmentation into small firms that can 'impede the integration of critical knowledge' (see Pisano 2006: 116). However, at the same time, the integration of knowledge capabilities in larger firms could prove detrimental in that it reduces the necessity for inter-organizational interaction and the attendant learning that such 'open innovation' entails (Chesbrough 2003).

Second, the complementary capabilities needed by life sciences firms means that the regional institutional ecosystem, especially the public science base, is a necessary, if not sufficient, reason for industrial 'clustering'(Woiceshyn 1995; Acharya *et al.* 1998; Cooke 2003; Feldman and Francis 2003). In contrast to the concentration of life sciences firms, the UK public science base was more evenly distributed in the early 2000s: there were 255 life sciences-related university departments (e.g. biology, chemistry, pharmacology etc.) with an average of 7 per NUTS2 region.[2] However, it is important to consider both the strength of these departments (see Zucker *et al.* 1998; Zucker *et al.* 2002) and the relative levels of public funding. Three of the four concentrations had a similar proportion of 'star' departments (i.e. rated 5* in the 2001 RAE) – 21–24 per cent – compared with 38 per cent of East Anglia's departments. This difference was reinforced when considering the extent of UK research council (RC) funding relevant to the life sciences: that is, funding by MRC, BBSRC and Natural and Environmental Research Council (NERC). The average regional funding from these three councils was around £19 million, although the relative funding per department shows that East Anglia received £6.54 million per department compared with around £4 million for Berks *et al.* and Eastern Scotland and only £2.22 million for Inner London. It is therefore possible to argue that East Anglia had a significantly stronger public science base than the other three concentrations; this does not necessarily mean that the life sciences sector in this region was stronger *per se*, rather that it was more likely to be dependent on university activity than elsewhere.[3]

Third, knowledge spillovers reinforce the importance of the regional institutional ecosystem because they provide an insight into the advantage that concentrated life sciences firms might have (see Audretsch and Feldman 1996). However, it is important to differentiate between different types of knowledge by distinguishing between appropriable (i.e. patents, which restrict access) and non-appropriable (i.e. publications, which are free to access) forms. Furthermore, patents can be split between more local (i.e. European Patent Office) and more international (i.e. US Patent Office) forms. Life sciences firms in the UK had produced 944 US patents and 804 European patents along with 4,209 journal publications. The NUTS2 regional averages were 26, 22 and 114 respectively. All three types of knowledge stock were regionally concentrated with half of US patents in just two regions (Berks *et al.* and Inner London), and European patents and articles concentrated in five regions. As this data showed, Berks *et al.* and Inner London had the most appropriable knowledge stocks, whilst Berks *et al.* and East Anglia had the most non-appropriable. This would suggest that knowledge spillovers in Berks *et al.* and East Anglia were more likely because they had more freely available knowledge (i.e. publications), although it is also significant that

East Anglia had far fewer patents than either Berks *et al.* and Inner London. The latter are more likely to have commercial potential and therefore be more relevant for life sciences innovation, although there is also a possibility that the appropriation of knowledge through patenting actually inhibits spillovers (Zeller 2008).

Finally, it is important to consider the extent of company alliances to illustrate both the formal relationships between firms and the local–global linkages in biotech innovation (Gertler and Levitte 2005; Birch 2008). As existing literature implies, multi-scale interactions are likely as important (if not more so) than local or cluster interactions (e.g. Malmberg and Power 2005). This is evident when considering the extent of the formal alliances of UK life sciences firms between 1997 and 2004. During this period, there were a total of 671 alliances involving UK firms with a regional average of 18. However, 70 per cent of all these alliances were limited to just three regions; East Anglia, Berks *et al.* and London. When considering the difference between national (n = 152) and international (n = 519) alliances, it is evident that there was a concentration of the latter in the same three regions (73 per cent) compared with a lower concentration of national alliances (60 per cent). These data illustrate the extent to which international linkages dominated the UK life sciences, especially in the main concentrations. Thus it is possible to argue that the main UK life sciences concentrations were tied into global alliance networks to a greater extent than other regions, although this may be the consequence of the location of specific large firms in those regions: for example, Celltech (bought by UCB in 2004) in Berks *et al.* and Cambridge Antibody Technology (bought by Astra Zeneca in 2006) in East Anglia.

The knowledge–space dynamic in the UK life sciences

While it is helpful to look at broad trends and patterns in the UK life sciences, this does not actually help us unpack what underpins innovation processes in the sector. As I argued above and in Chapter 2, it is necessary to look at the knowledge–space dynamic in order to analyze these processes. In particular, it is important to avoid the assumption that when firms, universities and other organizations are in close spatial proximity it becomes easier to transfer knowledge across organizational boundaries. This, according to Malmberg and Maskell (2002), reverses the causal explanation that knowledge is a critical element to regional innovation systems. Such assumptions are based on the notion that innovation processes result from regional knowledge capacity, especially when it comes to tacit knowledge, in contrast to actually showing how regional knowledge capacity drives innovation processes. To explore this issue in more depth, it is necessary to go beyond trends and patterns, which is where the structured interview data comes into play.

The structured interviews were designed to elucidate the frequency of contact that social actors in the life sciences innovation ecosystem – life sciences firms and service providers – drew upon three different forms of knowledge: codified, tacit, and commercial knowledge. I used ‘written forms’ as a proxy for codified knowledge and ‘face-to-face forms’ for tacit knowledge. Informants (n = 109) were asked to rate frequency of ‘contact’ with all three knowledge forms on a

Likert scale from 1 (no contact) to 5 (frequent contact), by sources (e.g. competitor, supplier, customer) and geographical scale (e.g. local, national, international). I split the findings into different categories, primarily between 'clustered' and 'non-clustered' where the former included any informants within the four life sciences concentrations I outline above.

Overall, the findings confirm the previous work of scholars like Breschi *et al.* (2001) and Leibovitz (2004) who argue that local knowledge and linkages are not as crucial to the life sciences as for other sectors. However, it also contrasts with research that stresses the link between local knowledge and global knowledge in specific 'nodes of excellence' (e.g. Coenen *et al.* 2004) because there are negative associations between both local and international explicit and tacit knowledge sources. Moreover, the findings contradict a number of assumptions about the spatial location of knowledge production and diffusion; i.e. that tacit knowledge is, by necessity, contextually specific and therefore difficult to transfer or acquire, whilst codified or explicit knowledge is 'ubiquitous' and therefore easier to transfer or acquire (Markusen 1996; Malmberg and Maskell 2002; Malecki 2000; Gertler 2003).

The findings illustrate the importance of analyzing the knowledge specificities of different places; for example, the data shows that no two places are alike. Consequently, it is important to analyze the different relationships between knowledge forms, sources and origins across the four life sciences concentrations: East Anglia, Inner London, Berks *et al.*, and Eastern Scotland. The geographical origin of different knowledge forms for the non-clustered and four clustered informants is shown in Figures 3.2 (codified), 3.3 (tacit), and 3.4 (commercial). These data are expressed in terms of the mean rating given by all informants regarding the source of different knowledge forms at different scales.

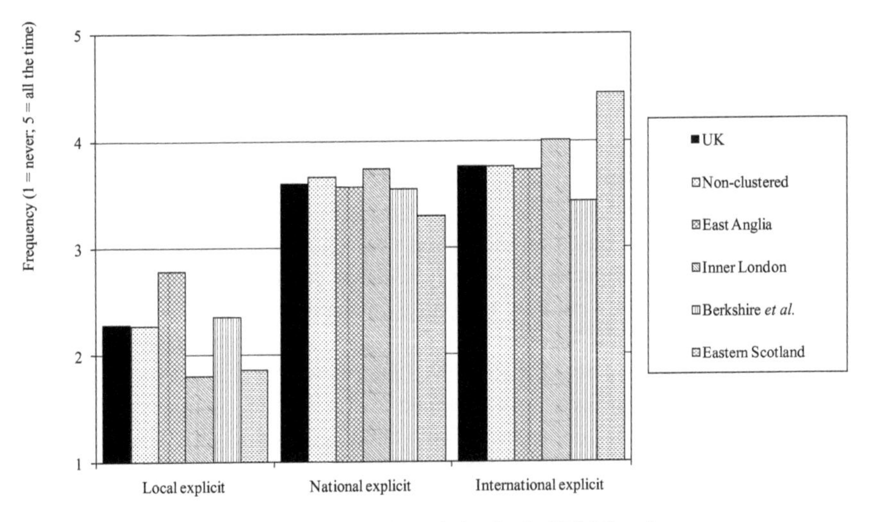

Figure 3.2 The 'clustering' of codified knowledge in the UK life sciences

Source: Birch (2006).

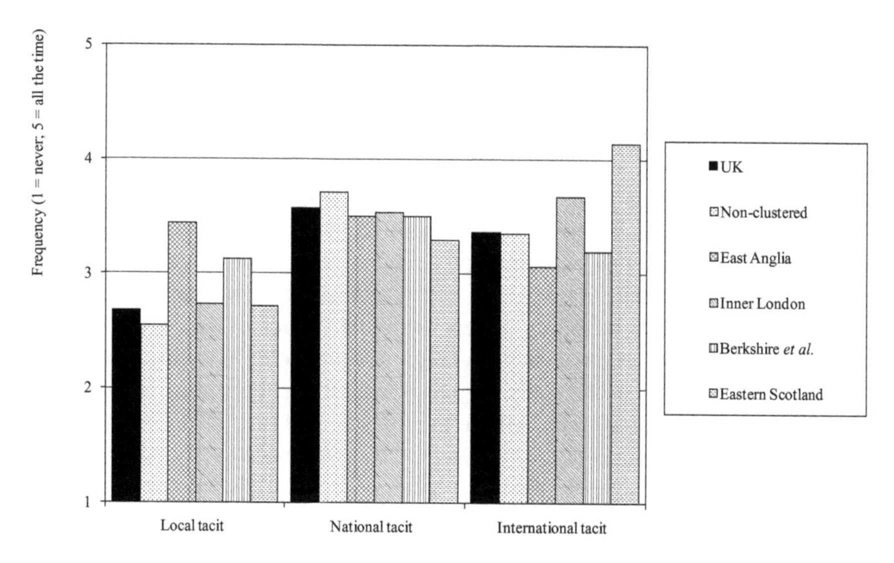

Figure 3.3 The 'clustering' of tacit knowledge in the UK life sciences
Source: Birch 2006.

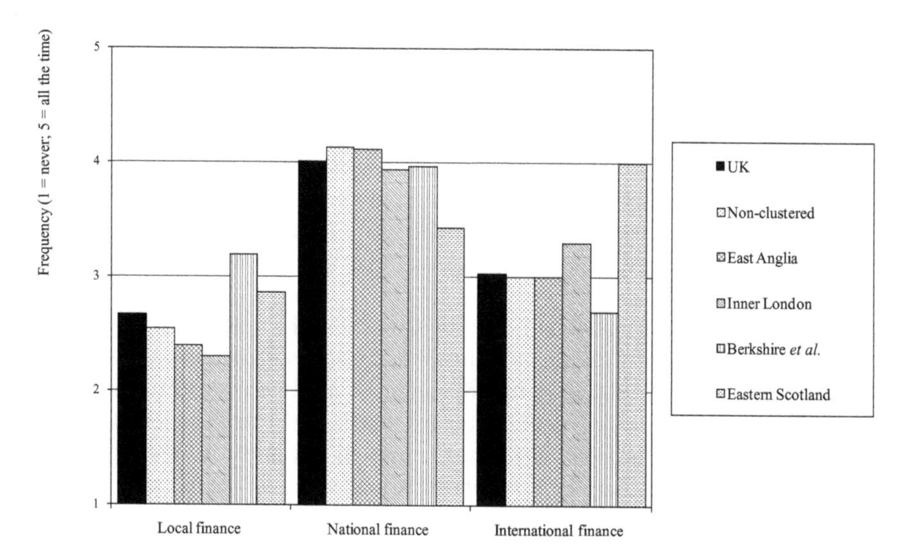

Figure 3.4 The 'clustering' of commercial knowledge in the UK life sciences
Source: Birch (2006).

As Figures 3.2 and 3.3 show, for example, East Anglia informants emphasized local codified and local tacit sources more than non-clustered informants. Moreover, they have less frequent contact with international tacit sources. The same trend is evident with Berks *et al.* informants who also had more contact with local tacit sources, as well as local commercial knowledge. In contrast, Inner London and Eastern Scotland informants were much more internationally focused with higher ratings for international explicit, tacit and commercial knowledge sources than non-clustered respondents (and the other concentrations) – see Figures 3.2, 3.3 and 3.4.

As this data shows, even supposedly clustered social actors draw from beyond their local or regional innovation ecosystem. Although informants drew from knowledge from their locality – or what can be theorized as their 'cluster' – it is notable that in many cases informants drew more on knowledge from *outwith* their cluster than within. This was the case whether it was codified, tacit or commercial knowledge. This finding largely contradicts much of the academic work and policy prescriptions on clusters and regional innovation systems, which have stressed the importance of locally-clustered institutions, dynamics, and interdependencies to innovation, especially in the life sciences (e.g. DTI 1999a, 1999b; Cooke 2001, 2003). It does support, though, the claims made by Malmberg (2003) and Malmberg and Power (2005) that local or regional clusters or innovation systems are not actually characterized by significantly higher levels of local or regional cooperation or more local or regional linkages. These findings do, moreover, demonstrate the need to theorize international and/or global cooperation and linkages in life sciences innovation, supporting the work of people like Bathelt *et al.* (2004) on 'global pipelines'.

Conclusion

In considering the knowledge and spatial patterns and processes of the UK life sciences, it is important to understand the *knowledge–space dynamic* that produces geographical specificity in the different innovation systems of the four UK regional concentrations: East Anglia, Berks *et al.*, Inner London, and Eastern Scotland. This concept helps to explain why the UK life sciences sector is so unevenly spread.

The life sciences is embedded in particular regions because these regions are constituted by a *knowledge–space dynamic* that leads to geographically-specific innovation processes, although this dynamic is, necessarily, multi-scalar and should not be conceived as geographically localized. In this chapter, I sought to incorporate insights from different theories to develop a concept that can account for this specificity by considering knowledge *and* spatial patterns and processes, rather than one or the other. It has shown how the UK life sciences is concentrated in certain regions that have different and varied elements that all contribute to the innovation ecosystem; for example, differences in public science base, knowledge spillovers, extent and size of biotech firms etc. More importantly, however, is that the chapter has shown how knowledge and innovation are multi-scalar processes. Social actors do not simply draw on local spillovers or regional institutional assets, and such like. Rather, social actors draw on a range of knowledge forms, from different sources, at different geographical scales (Birch 2008). It is therefore crucial to consider how life sciences innovation is reliant on *multi-scale* processes and linkages that go beyond a 'local–global' binary (e.g. Bathelt *et al.* 2004) and the emphasis on localized learning (see Malmberg and Power 2005).

A final issue to address is what the uneven spread of the UK life sciences means for regional development. Although I focus on this issue in Chapter 7, it is helpful to raise some concerns here. It would appear that the promotion of life sciences *clusters* by national and regional policy-makers (e.g. Scottish Enterprise, DTI) will not necessarily be successful considering the contributory components in life sciences innovation. In particular, the public science base plays an important role, yet it is also subject to national (and supranational) policy initiatives to create centres of excellence. Such centres may, paradoxically, inhibit innovation in that they limit the options available to an industry dependent upon multi-scalar knowledge capabilities by embedding public and private investment in a small number of locations. This could also encourage a parochial approach to innovation through focusing attention only on this limited number of sites of innovation. Finally, the focus on high technology sectors like the life sciences ignores the limited size of the sector itself. A few hundred firms and few thousand jobs would not seem to offer a viable source of economic growth, let alone regional development in places that have suffered most from deindustrialization, and necessitate investment in other areas as well.

Notes

1 The NUTS scale is used by the European Union (EU) to designate statistical regions across Europe and refers to regions with populations between 800,000 and 3 million (http://ec.europa.eu/eurostat/ramon/nuts/home_regions_en.html; accessed June 2007).
2 See DTI (1999b) for list of university departments identified as relating to the biotech industry.
3 It is also possible to look at the extent of service providers (e.g. venture capital, business angels, management consultants, lawyers, accountants etc.), which reveals a greater concentration of such complementary capabilities in Inner London with a third of all service providers in the UK. Consequently, it is possible to argue that although such 'complicit' knowledge (Cooke 2007) are important – in that all the four concentrations have above average numbers of service providers – they are again not sufficient in themselves.

References

ACARD, ABRC and The Royal Society (1980) *Biotechnology: Report of a Joint Working Party* [aka *The Spinks Report*], HMSO: London.

Acharya, R., Arundel, A. and Orsenigo, L. (1998) The evolution of European biotechnology and its future competitiveness, in J. Senker (ed.) *Biotechnology and Competitive Advantage*, Cheltenham: Edward Elgar.

Acs, Z., Audretsch, D. and Feldman, M. (1991) Real effects of academic research: Comment, *American Economic Review* 82: 363–367.

Asheim, B. and Coenen, L. (2006) The role of regional innovation systems in a globalising economy, in G. Vertova (ed.) *The Economic Geography of Globalization*, London: Routledge.

Audretsch, D. and Feldman, M. (1996) R&D spillovers and the geography of innovation and production, *The American Economic Review* 86: 630–640.
Audretsch, D. and Stephan, P. (1996) Company–scientist locational links: The case of biotechnology, *The American Economic Review* 86: 641–652.
Audretsch, D. and Stephan, P. (1999) Knowledge spillovers in biotechnology: sources and incentives, *Journal of Evolutionary Economics* 9: 97–107.
Bagchi-Sen, S. (2007) Strategic considerations for innovation and commercialization in the US biotechnology sector, *European Planning Studies* 15: 753–766.
Bartholomew, S. (1996) National systems of biotechnology innovation: Complex interdependence in the global system, *Journal of International Business Studies* 28: 241–266.
Bathelt, H., Malmberg, A. and Maskell, P. (2004) Clusters and knowledge: Local buzz, global pipelines and the process of knowledge creation, *Progress in Human Geography* 28: 31–56.
BIGT (2003) *Improving National Health, Improving National Wealth*, London: Bioscience Innovation and Growth Team.
Birch, K. (2006) *Biotechnology Value Chains as a Case Study of the Knowledge Economy: The Relationship between Knowledge, Space and Technology*, Unpublished PhD thesis, Department of Planning, Oxford Brookes University.
Birch, K. (2007a) Knowledge, space and biotechnology, *Geography Compass* 1: 1097–1117.
Birch, K. (2007b) The social construction of the biotech industry, in P. Glasner, P. Atkinson and H. Greenslade (eds) *New Genetics, New Social Formations*, London: Routledge, pp. 94–113.
Birch, K. (2008) Alliance-driven governance: Applying a global commodity chains approach to the UK biotechnology industry, *Economic Geography* 84: 83–103.
Birch, K. (2009) The knowledge–space dynamic in the UK bioeconomy, *Area* 41(3): 273–284.
Birch, K. and Cumbers, A. (2010) Knowledge, space and economic governance: The implications of knowledge-based commodity chains for less-favoured regions, *Environment and Planning A* 42(11): 2581–2601.
Braczyk, H.-J., Cooke, P. and Heidenreich, M. (eds) (1998) *Regional Innovation Systems*, London: UCL Press.
Breschi, S., Lissoni, F. and Orsenigo, L. (2001) Success and failure in the development of biotechnology clusters: The case of Lombardy, in G. Fuchs (ed.) *Comparing the Development of Biotechnology Clusters*, London: Harwood Academic Publishers.
Bud, R. (1993) *The Uses of Life: A History of Biotechnology*, Cambridge, UK: Cambridge University Press.
Chakrabarti, A. and Weisenfeld, U. (1991) An empirical analysis of innovation strategies of biotechnology firms in the US, *Journal of Engineering and Technology Management* 8: 243–260.
Chesbrough, H. (2003) *Open Innovation*, Cambridge MA: Harvard Business School Press.
Coenen, L., Moodysson, J. and Asheim, B. (2004) Nodes, networks and proximities: On the knowledge dynamics of the Medicon Valley biotech cluster, *European Planning Studies* 12: 1003–1018.
Cooke, P. (2001) Biotechnology clusters in the UK: Lessons from localisation in the commercialisation of science, *Small Business Economics* 17: 43–59.
Cooke, P. (2003) The evolution of biotechnology in three continents: Schumpeterian or Penrosian?, *European Planning Studies* 11: 757–763.
Cooke, P. (2004a) The molecular biology revolution and the rise of bisocience megacentres in North America and Europe, *Environment and Planning C* 22: 161–177.

Cooke, P. (2004b) Regional knowledge capabilities, embeddedness of firms and industry organisation: Bioscience megacentres and economic geography, *European Planning Studies* 12: 625–641.
Cooke, P. (2005) Regionally asymmetric knowledge capabilities and open innovation: Exploring 'Globalisation 2' – A new model of industry organisation, *Research Policy* 34: 1128–1149.
Cooke, P. (2006) Global bioregions: Knowledge domains, capabilities and innovation system networks, *Industry and Innovation* 13: 437–458.
Cooke, P. (2007) *Growth Cultures*, London: Routledge.
Coriat, B., Orsi, F. and Weinstein, O. (2003) Does biotech reflect a new science-based innovation regime?, *Industry and Innovation* 10: 231–253.
Deeds, D. and Hill, C. (1996) Strategic alliances and the rate of new product development: An empirical study of entrepreneurial biotechnology firms, *Journal of Business Venturing* 11: 41–55.
DTI (1999a) *Biotechnology Clusters Report*, London: Department of Trade and Industry.
DTI (1999b) *Genome Valley Report*, London: Department of Trade and Industry.
Etzkowitz, H. and Leydesdorff, L. (2000) The dynamics of innovation: From National Systems and 'Mode 2' to a Triple Helix of university–industry–government relations, *Research Policy* 29: 109–123.
Feldman, M. (1999) The new economics of innovation, spillovers and agglomeration: A review of empirical studies, *Economic Innovation and New Technology* 8: 5–25.
Feldman, M. (2003) The locational dynamics of the US biotech industry: Knowledge externalities and the anchor hypothesis, *Industry and Innovation* 10: 311–328.
Feldman, M. and Francis, J. (2003) Fortune favours the prepared region: The case of entrepreneurship in the Capitol region biotechnology cluster, *European Planning Studies* 11: 765–788.
Gertler, M. (2003) Tacit knowledge and the economic geography of context, or The undefinable tacitness of being (there), *Journal of Economic Geography* 3: 75–99.
Gertler, M. and Levitte, Y. (2005) Local nodes in global networks: The geography of knowledge flows in biotechnology innovation, *Industry and Innovation* 12: 487–507.
Gottweis, H. (1998a) *Governing Molecules*, London: MIT Press.
Gottweis, H. (1998b) The political economy of British biotechnology, in A. Thackray (ed.) *Private Science*, Philadelphia: University of Pennsylvania Press.
House of Lords (1993) *Select Committee on Science and Technology: Regulation of the United Kingdom Biotechnology Industry and Global Competitiveness*, London: HMSO.
Howells, J. (2000) Knowledge, innovation and location, in J. Bryson, P. Daniels, N. Henry and J. Pollard (eds) *Knowledge, Space, Economy*, London: Routledge.
Howells, J. (2002) Tacit knowledge, innovation and economic geography, *Urban Studies* 39: 871–884.
Kaiser, R. (2003) Multi-level science policy and regional innovation: The case of the Munich cluster for pharmaceutical biotechnology, *European Planning Studies* 11: 841–857.
Kasabov, E. and Delbridge, R. (2008) Innovation, embeddedness and policy: Evidence from life sciences in three UK regions, *Technology Analysis & Strategic Management* 20: 185–200.
Malecki, E. (1997) *Technology and Economic Development*, Essex: Longman.
Malecki, E. (2000) Creating and sustaining competitiveness: Local knowledge and economic geography, in J. Bryson, P. Daniels, N. Henry and J. Pollard (eds.) *Knowledge, Space, Economy*, London: Routledge.

Malmberg, A. (2003) Beyond the cluster – Local milieus and global connections, in J. Peck and H. Yeung (eds) *Remaking the Global Economy*, London: SAGE.

Malmberg, A. and Maskell, P. (2002) The elusive concept of localization economies: Towards a knowledge-based theory of spatial clustering, *Environment and Planning A* 34: 429–449.

Malmberg, A. and Power, D. (2005) (How) do (firms in) clusters create knowledge?, *Industry and Innovation* 12: 409–431.

Markusen, A. (1996) Sticky places in slippery space: A typology of industrial districts, *Economic Geography* 72(3): 293–313.

Mirowski, P. (2011) *ScienceMart*, Cambridge, MA: Harvard University Press.

Mittra, J. (2016) *The New Health Bioeconomy: R&D Policy and Innovation for the Twenty-first Century*, Basingstoke: Palgrave Macmillan.

Moodysson, J., Coenen, L. and Asheim, B. (2008) Explaining spatial patterns of innovation: Analytical and synthetic modes of knowledge creation in the Medicon Valley life-science cluster, *Environment and Planning A* 40: 1040–1056.

ODPM (2004) *Our Towns and Cities: The Future*, London: Office of the Deputy Prime Minister.

Owen, G. (2001) *Entrepreneurship in UK Biotechnology: The Role of Public Policy*, Working Paper No. 14: The Diebold Institute.

Owen, G. and Hopkins, M. (2016) *Science, the State and the City*, Oxford: Oxford University Press.

Pisano, G. (2006) Can science be a business? Lessons from biotech, *Harvard Business Review* 84: 114–125.

Polanyi, M. (1966) *The Tacit Dimension*, New York: Doubleday.

Porter, M. (1990) *The Competitive Advantage of Nations*, London: Macmillan.

Powell, W. (1998) Learning from collaboration: knowledge and networks in the biotechnology and pharmaceutical industries, *California Management Review* 40: 228–240.

Powell, W., Koput, K. and Smith-Doerr, L. (1996) Interorganizational collaboration and the locus of innovation: Networks of learning in biotechnology, *Administrative Science Quarterly* 41: 116–145.

Powell, W., Koput, K., Bowie, J. and Smith-Doerr, L. (2002) The spatial clustering of science and capital: Accounting for biotech firm–venture capital relationships, *Regional Studies* 36: 291–305.

Prevezer, M. (2003) The development of biotechnology clusters in the USA from the late 1970s to the early 1990s, in G. Fuchs (ed.) *Biotechnology in Comparative Perspective*, London: Routledge.

Ryan, C. and Phillips, P. (2004) Knowledge management in advanced technology industries: An examination of international agricultural biotechnology clusters, *Environment and Planning C* 22: 217–232.

Saviotti, P., Joly, P.-B., Estades, J., Ramani, S. and de Looze, M.-A. (1998) The creation of European dedicated biotechnology firms, in J. Senker (ed.) *Biotechnology and Competitive Advantage*, Cheltenham: Edward Elgar.

Senker, J. (2005) *Biotechnology Alliances in the European Pharmaceutical Industry: Past, Present and Future*, SPRU, University of Sussex: SEWPS Paper No. 137.

Senker, J., Joly, P.-B. and Reinhard, M. (1996) *Overseas Biotechnology Research by Europe's Chemical/Pharmaceutical Multinationals: Rationale and Implications*, SPRU, University of Sussex: STEEP Discussion Paper No. 33.

Sharp, M. (1985) *The New Biotechnology: European Governments in Search of a Strategy*, University of Sussex: Sussex European Papers No.15.

Woiceshyn, J. (1995) Lessons in product innovation: A case study of biotechnology firms, *R&D Management* 25: 395–409.

Zeller, C. (2001) Clustering biotech: A recipe for success? Spatial patterns of growth of biotechnology in Munich, Rhineland and Hamburg, *Small Business Economics* 17: 123–141.

Zeller, C. (2004) North Atlantic innovative relations of Swiss pharmaceuticals and the proximities with regional biotech arenas, *Economic Geography* 80: 83–111.

Zeller, C. (2008) From the gene to the globe: Extracting rents based on intellectual property monopolies, *Review of International Political Economy* 15: 86–115.

Zucker, L., Darby, M. and Armstrong, J. (2002) Commercializing knowledge: University science, knowledge capture, and firm performance in biotechnology, *Management Science* 48: 138–153.

Zucker, L., Darby, M. and Brewer, M. (1998) Intellectual human capital and the birth of US biotechnology enterprises, *The American Economic Review* 88: 291–306.

4 Innovation governance in the Scottish life sciences

Co-authored with Andrew Cumbers, University of Glasgow

Introduction

As the last chapter illustrated, innovation is a multi-scalar process, meaning that overly localized or regionalized analytical perspectives (e.g. clusters) can miss elements that are central to an understanding of the geography of innovation This is especially the case when it comes to new sectors, like the life sciences, that are dependent on new forms of innovation governance because research and commercialization are so geographically distributed and dispersed, globally as well as nationally. It is important, in this context, to explore how (regional) social actors can manage and benefit from their integration into global commodity chains. My aim in this chapter, then, is to address these issues by examining the positioning of the life sciences in Scotland using a knowledge-based commodity chains perspective in order to understand better the different forms of innovation governance at play in a 'less-favoured region' (LFR). This provides one way to understand how other LFRs can try to capture value from new high-technology sectors and upgrade their regional capacities.

As mentioned in Chapter 2, one response countries have taken in the Global North to the competitive threats arising from globalization and the shift of more basic forms of manufacturing and services to the Global South is to promote the idea of 'competitiveness' through the expansion of knowledge-based industries – such as the life sciences – in which innovation drives production and capital expansion (Reich 1991). Mainstream debates have centred on several issues with regards to this so-called *knowledge-based economy* (KBE), including: the growing importance of scientific research as a source of innovation and new commodities; the growth in knowledge intensity as knowledge replaces other factors of production; the need for organizational change to facilitate learning and the capture of knowledge in new commodity forms; and, the central place of universities and education in economic development (see Kitagawa 2004; Powell and Snellman 2004; see Warhurst and Thompson 2006 for a critical review).

The KBE agenda has come to dominate social and economic policy in several economies, especially in the EU where the *Lisbon Agenda* explicitly linked global competitiveness to the continuation of the European 'social model' (e.g. European Council 2000) – see Chapter 5. More critically, several scholars have queried whether the reshaping brought about by this new political-economic project will actually benefit all places equally, or whether it is more likely to accentuate continuing processes of capital accumulation, prioritizing certain types of knowledge and places over others (see Jessop 2006; Birch and Mykhnenko 2009; Hudson 2011). What this new policy vision and strategy means for LFRs is less clear, especially given the continuing evidence of disparities in economic experience between different regions (Birch *et al.* 2010). For example, it is not clear whether the pursuit of the KBE agenda has or will alleviate uneven development between regions, since LFRs may lack the social and economic infrastructure needed to attract and embed new forms of knowledge-based employment. Part of the answer to this sort of question involves understanding how the KBE recasts relations between regions and firms in knowledge-based commodity chains, implying new possibilities for LFRs (Morgan 1997; Hudson 1999; MacKinnon *et al.* 2002). Are power differentials and flows of value between actors and regions within knowledge-based commodity chains different to other sectors and what are the governance implications for innovation of these differences? All these issues are centrally implicated in the need to look again at the processes that underpin regional development in knowledge-based sectors, especially in relation to changing geographies of innovation governance.

In this chapter, I engage with these debates through a case study of the life sciences industry in Scotland, which I define as an LFR. The KBE discourse has been particularly important in Scotland where, since the advent of devolved government in 1999, it has been central to regional policy formation and governance (Rosiello 2004). Nowhere has this been more prevalent than in the country's rapidly developing life sciences industry, a regional success story against a backdrop of long-term decline in traditional manufacturing sectors and failed regional policy initiatives. The ability to continue its development in the future hinges upon the ways that Scottish life science firms are plugged into the broader global geography of the life sciences. Particularly critical in this respect are the mechanisms of coordination, governance and power that are emerging in life science commodity chains. I start by briefly outlining the analytical approach, already discussed in Chapter 2, which draws upon recent literature on GCCs and builds upon the conceptualization of these new geographies of knowledge production in the life sciences as dependent upon *alliance-driven governance* (Birch 2008). Following this I briefly outline the research context and methodology before presenting the empirical findings on the forms of innovation governance emerging between actors in the life sciences sector.

Innovation governance in knowledge-based commodity chains

Alliance-driven governance

Over the recent past, as the limits to clustering and agglomeration have been identified (MacKinnon *et al.* 2002; Malmberg and Power 2005; Asheim *et al.* 2006), there has been greater recognition of the importance of extra-local linkages, particularly linkages that tie regional systems into broader global networks (e.g. Cumbers *et al.* 2003; Bathelt *et al.* 2004; Moodysson 2008). In the life sciences, this research has, however, often emphasized the concentration of 'whole value chains' in certain places (Coenen *et al.* 2004; Cooke 2007), only partially addressing the diverse geographical patterns, processes and governance of life science innovation captured by the notion of 'local nodes' in 'global networks' (e.g. Gertler and Levitte 2005; Gertler and Vinodrai 2009). It is crucial to consider these geographical dimensions in more depth because of the different constituents of innovation governance in the life sciences, ranging from regional knowledge capacities and infrastructure through national regulatory systems to global trade rules. Consequently, a focus on localized interaction and relationships misses the concentration *and* dispersal of innovation processes and governance across multiple scales and different places (Malmberg and Power 2005). In this sense, innovation involves 'linkages and interrelationships *between* and *across* these various spatial levels or scales' (Bunnell and Coe 2001: 577; also Coe and Bunnell 2003), which means that any approach that places too much emphasis on the 'locally-boundedness' (Phelps 2004) of such processes and governance misses the diverse spatialities of innovation governance.

In order to understand the geographies of innovation in the life sciences, there is therefore a need to go beyond localized interaction and learning to highlight the importance of knowledge-generating processes at a range of different geographical scales. While there has been limited work in this regard, some analyses have drawn on the global commodity chains (GCC) perspective (Birch 2008; Haakonsson 2009). The main conceptual benefit this approach provides is that it avoids both a firm-centric and region-centric focus and thereby enables an examination of the diverse geographical linkages and relationships that make up life sciences innovation, production and governance.

The GCC approach has its origins in world-systems theory (Hopkins and Wallerstein 1986), but is most identified with the later work of Gary Gereffi (1994, 1996). Whilst Gereffi has since moved towards a global value chain (GVC) approach (see Gereffi *et al.* 2005), this latter conceptualization focuses more explicitly on the relationship between two actors within a chain and, therefore, it provides a limited purchase on the multi-scalar context in which such actors operate (Bair 2005; Haakonsson 2009). While the GCC approach is a useful starting point, it does however have its limitations for exploring the more complex and multi-centred economic relations that constitute knowledge-based commodity chains. For example, the focus on lead firms (e.g. dominant manufacturing companies or retail chains) that drive the value chain is problematic when it comes to knowledge-based sectors where it is more difficult for dominant actors to control processes of value capture because of the uncertainties and complexities involved in knowledge production (Birch 2008). Where knowledge is considered in these GCC models it is relatively unproblematic within asymmetrical power relations in which the task for host regions engaging with corporate actors is to upgrade their knowledge intensive activities, typically through knowledge transfer from more developed regional assemblages, in a way that might sustain and enhance value capture.

In attempting to advance our understanding of innovation coordination and governance in knowledge-based commodity chains, I conceptualized an *alliance-driven governance* (ADG) model that reflects the flatter and less hierarchical networks of knowledge and production processes through which such activities operate in the life sciences (Birch 2008). Unlike the earlier GCC models, this concept reflects the essentially non-linear realities of knowledge production such that traditional regional development concerns with upgrading are replaced by attention to the changing dynamics of regional embeddedness within global knowledge networks. In this sense it accounts for two important analytical points that are currently under-theorized in the GCC literature. First, the relative 'stickiness' of knowledge in different places (Markusen 1996), which can provide the locational assets necessary to withstand or ameliorate the worst impacts of industrial restructuring. Second, the idea of regional development and restructuring as more than the local capabilities of organizations and institutions; it also consists of the interaction between local and extra-local actors and new priorities and agendas that arise across different geographies.

The ADG model consists of a number of characteristics particular to the coordination and governance of new high-technology sectors like the life sciences (see Table 4.1). Specifically, these include the need for core competencies in collaborating with and acquiring and absorbing knowledge from diverse organizations (Senker 2005), as well as the capability to operate across different and distinctive institutional and regulatory regimes in order to access multiple markets (Ossenbrugge and Zeller 2002). All this entails both an inherent uncertainty and the coordination of diverse incentives, which necessitates a reliance on certain forms of capital investment that is neither short-term nor risk-averse; for example, public and venture capital (Casper and Kettler 2001). Such uncertainty is the consequence of the high asset specificity of new scientific and technological discoveries and innovations whose value can be both initially unclear and intangible (e.g. intellectual property) and entail long-term development costs (i.e. high opportunity costs), necessitating new forms of (intellectual) property protection to encourage investment (Arora and Merges 2004). Consequently, it is evident that such a model is built around new geographical forms of innovation governance in which firms and other organizations engage in numerous relationships and collaborations with a variety of different actors ranging from publicly-funded universities through multi-national corporations to government regulatory agencies.

Innovation governance and trust in knowledge-based commodity chains

The nature of trust and its spatial implications are central to innovation governance. From the perspective of social actors, trust can be defined 'as the judgment one makes on the basis of one's past interactions with others that they will seek to act in ways that favour one's interests, rather than harm them, in circumstances that remain to be defined' (Lorenz 1999: 305). In existing research on innovation and regional development, trust has been viewed as an

Table 4.1 The alliance-driven governance in knowledge-based commodity chains

Model	*Characteristics*	*Theoretical underpinnings*	*Spatial implications*
GCC drivers	'Patient capital'	High asset specificity of new science and technology leads to risk and uncertainty and therefore discourages short-term, low-risk investment necessitating public funding and venture capital (VC).	Embedding of investment in the institutional arrangement of national and regional governments (and their agencies) and private capital (e.g. VC).
Core competencies	Collaborating, regulations	Specialisation and complexity preclude integration so organizations rely on collaborations that cross national regulatory regimes necessitating an understanding of different regulatory standards.	Dispersal of the innovation process between and across organizations in different regions and countries.
Entry barriers	Economies of complexity	High-cost, analytical knowledge (i.e. science) infrastructure and diverse national regulatory policies inhibit entry.	Concentration of scientific capacity and linkages within broader global knowledge pipelines.
Sectors	High-technology, intangibles	Sectors dependent upon intellectual property protection to ensure value capture and pursuit of regulatory arbitrage.	Dependence on national (e.g. patent offices) and global (e.g. WTO) governance institutions.
Network linkages	Alliance based	Requirements of collaborating and regulatory adherence mean that organizations rely on the coordination of diverse incentives.	Inter-organizational and cross-border knowledge, resource and regulatory linkages and coordination.
Network structure	Matrix	The collective nature of innovation and high asset specificity mean that networks consist of numerous interactions.	Multi-organizational, multi-scalar and dynamic knowledge networks.

Source: Birch and Cumbers (2010), reproduced with permission of SAGE.

important asset underpinning economic performance, especially in knowledge-based activities that require considerable interaction and collaboration between actors and where, departing from mainstream economic analysis, knowledge is always assumed to be partial (Storper 1997).

Discussions of trust have tended to focus upon it as a localized phenomenon, bound up in endogenous business and development networks, which when it operates successfully provides considerable 'untraded interdependencies' for regions and small firms in local business clusters (Storper 1995; Cooke and Morgan 1998). In these circumstances, researchers have found it useful to distinguish between 'competence trust', where collaborators can be confident (usually through reputation and past performance) that their collaborators have the ability to carry out certain activities to a required standard, and 'goodwill trust' which refers to a deeper level of trust, often tied up in personal or social relations that go beyond a narrowly defined economic dividend and are present in longer-term networks and associations (Lazaric and Lorenz 1998; MacKinnon *et al.* 2004).

As is now widely recognized in the literature, trust is not a localized phenomenon but is critical to the governance of global business networks, especially in knowledge-intensive sectors (MacKinnon *et al.* 2004). In alliance-driven governance (ADG) the intersection of diverse communities of practice requires both kinds of trust, developed through more open networks of knowledge exchange and intellectual endeavour, alongside the need for reliability in delivering a product to the marketplace. Tensions necessarily arise in these trust relations between the need for openness in the exchange of ideas and the economic imperatives to capture value through the protection and definition of intellectual property rights. Furthermore, the complexity of knowledge-based commodity chains, as in the life sciences, entails a further dimension of trust that is related to the need to meet particular regulatory standards and protocols (e.g. International Standards Organisation). This involves the *objectification* of relationships, where it is the adherence to standards themselves that engenders trust rather than the relationship itself. Here we conceptualize this as *objective trust* standing in contrast to the inter-subjective forms of trust outlined above. Such objective trust is embodied in objects and practices that adhere to particular operational standards (e.g. clinical, manufacturing), which ensures that the trustworthiness (i.e. 'quality') of those objects and practices is bound up with the process of standard-setting rather than previous performance or personal connections.

The potential implications of trust in alliance-driven governance are profound for LFRs for they suggest that knowledge-based commodity chains may be subject to very different imperatives to those that structure spatial divisions of labour between regions in other sectors (Massey 1995). Successful product development requires collaboration and the bringing together of diverse and complex knowledges across organizations and space in a dynamic and uncertain marketplace that defies attempts to dominate and centralize power but instead requires the exercise of very different spatial modalities of governance (Allen 2003) to facilitate the kinds of collaboration and trust identified here. In particular, alliance relations require the development of more associational forms of governance and trust, on the one hand, with the kind of weaker ties identified by Grabher (1993), on the other hand, for the formation of fluid and temporary assemblages through which successful commodification takes place.

For LFRs, outside the core high-technology centres, opportunities arise to engage with these wider and more open global knowledge networks and even reposition themselves within broader spatial divisions of labour and imply that there is potential for public policy intervention. Set against these possibilities however, existing processes of uneven development and their effects in terms of the unequal distribution of knowledge assets, such as research and development expertise, elite universities, access to finance and venture capital and positioning within broader global knowledge networks still tend to favour core regions.

Scotland and the life sciences

Methodological note

The empirical work I draw on in this chapter is based on a study of the Scottish life sciences carried out in 2008 using a multi-method research design incorporating a number of different stages. The first stage involved identifying the 600 organizations included in the life sciences using Scottish Enterprise's 'Sourcebook' database. The second stage involved a survey of all the 'core' life science firms, with 'core' defined as those firms using biological techniques and applications to develop products or intellectual property. Under this definition, it was possible to identify and map 190 firms from which to survey; the response rate was 39 per cent after telephone follow-ups ($n = 74$). Beyond mapping the local geography of the life sciences, the purpose of the survey was to position Scottish firms within the global life science commodity chain and to identify the types of backwards and forwards linkages that existed, and the geography of these connections (whether local or trans-local).

The third part of the research involved a series of 32 in-depth interviews with a range of social actors in the Scottish life sciences. The interviews were split between informants from life sciences firms and 'institutional' actors. The first set of interviews included 19 informants from 18 life science firms. These informants were managing directors or research directors of Scottish-owned firms (12 firms) and Scottish-based subsidiaries (6 firms). All the Scottish-owned firms were engaged in R&D, whilst the Scottish-based subsidiaries were mostly engaged in manufacturing operations although some also carried out R&D. The 'institutional' interviews included 13 informants drawn from various Scottish organizations connected to the life sciences (e.g. Scottish Government, trade associations, business angels, industry advisory bodies, etc.). The interviews provide the basis for the discussion in this chapter about the governance and coordination of activity along knowledge-based commodity chains, highlighting, in particular, how power plays out between different actors and the impacts that this has on regional economic development.

Positioning Scotland in life sciences commodity chains

The life sciences sector has been one of the few instances of economic success and renewal in Scotland against a broader background of economic decline in the country's traditional manufacturing industries. Scotland represents an interesting case because it is a 'region' of the UK that has suffered from severe de-industrialization and uneven development over the last thirty years, losing around 60 per cent of its manufacturing employment since the late 1970s (Birch *et al.* 2010). Until recently, large parts of central and lowland Scotland were deemed worthy of considerable European regional assistance because of poor economic performance and social deprivation. It is therefore a 'region' that can be considered 'less-favoured', even though some places such as Edinburgh and Aberdeen have grown strongly over the last three decades.

As noted in Chapter 3, parts of Scotland – primarily around Edinburgh and Dundee – represent one of four key concentrations of life sciences innovation in the UK. In this chapter, however, I focus on Scotland as a whole. In 2008, the life sciences sector employed around 30,000 people across 600 private and public sector organizations (including the Universities of Aberdeen, Dundee, Edinburgh and Glasgow) making Scotland the most significant 'cluster' of life sciences activity in the UK outside London and the south east of England (Birch 2009). As Figure 4.1 illustrates, the life sciences, broadly speaking, is concentrated in the central belt of Scotland between the major cities of Glasgow, Edinburgh and Dundee, with smaller sub-clusters around Aberdeen and Inverness.

The emergence of the Scottish life sciences owes much to a long tradition of bio-medical research in Scottish universities (extending as far back as the 18th century Enlightenment), allied to public sector support for medical research dating back to the 1940s. Since the early 1990s, the main development agency Scottish Enterprise has also provided considerable support for local business start-ups and spin-off companies from university research as well as providing longer-term support for innovation through a number of initiatives (Rosiello 2007). Consequently, most of the 74 firms surveyed had been established in the last two decades; 39 per cent in the 2000s and another 33 per cent in the 1990s. This provides encouraging signs that LFRs like Scotland are not destined to lose out on the economic benefits of new knowledge-based industries.

The successful development of the life sciences has been based on sustained collaboration between key actors in government, universities and the business sector. While these regional (and national) institutional arrangements have been critical, the growth of the life sciences has been facilitated by wider spatial networks from the outset. Of particular importance are the global knowledge communities within which Scottish scientists and academics operate, and a considerable Scottish scientific diaspora which is reproduced by flows of labour and knowledge of varying durations to and from 'the region'. Added to this, the geography of the life sciences market, in terms of the considerable economies and technologies of scale and the subsequent dominance of large transnational corporations, means that Scottish firms must be 'global' from the outset in selling their products and ideas.

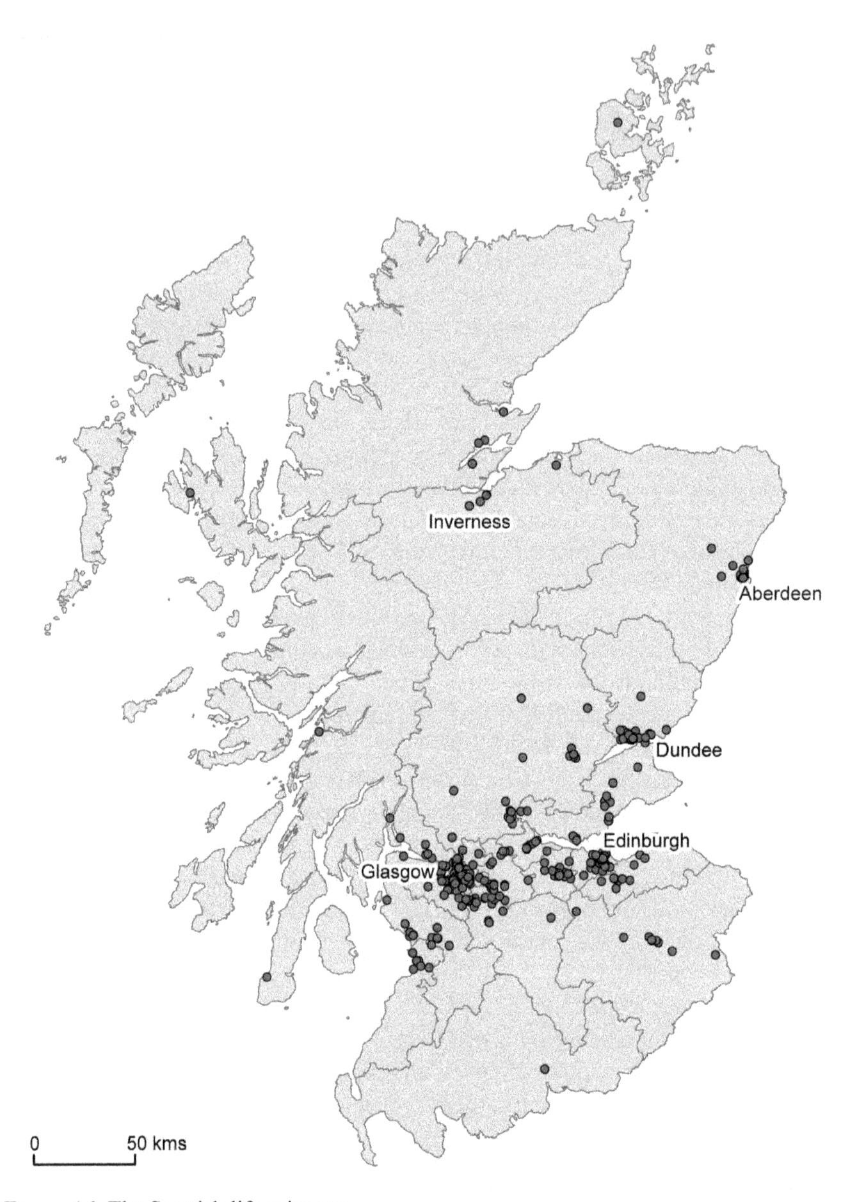

Figure 4.1 The Scottish life sciences

Note: data from *Scottish Enterprise Life Science Sourcebook* 2006/07; reproduced with permission of this book's author.

Knowledge-based commodity chains in a less-favoured region

I start the empirical analysis by looking at the geographies of knowledge-based commodity chains in the Scottish life sciences. Drawing on the results of the

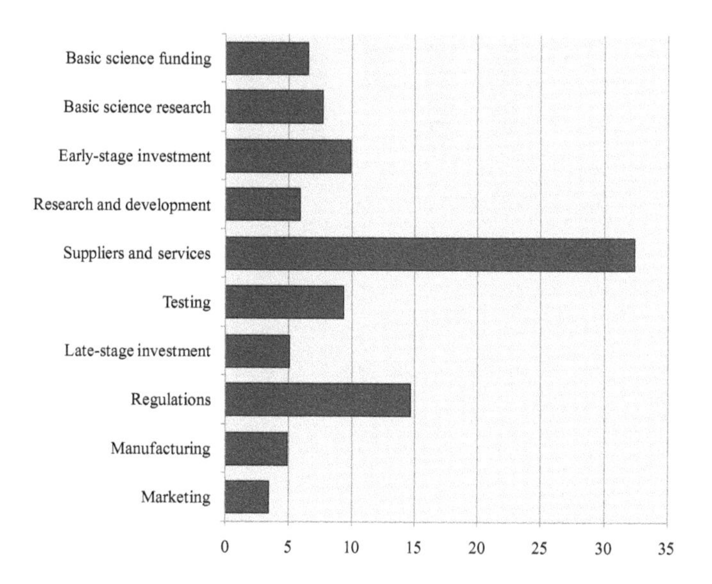

Figure 4.2 Commodity chain relationships in the Scottish life sciences

Source: Birch (2011), reproduced with permission of Wiley Blackwell.

survey I can show the multiple geographies implicated in life science innovation and how these reflect the position of Scottish firms in knowledge-based commodity chains. In considering these particular geographies, I focus on three main patterns: first, the number of relationships that Scottish life science firms have with other organizations along the commodity chain; second, the location of these other organizations; and third, the importance of the Scottish public sector along the commodity chain.

First, in looking at the total number of inter-organizational relationships that surveyed Scottish life science firms had (n = 898, mean = 12), it is evident that certain external linkages along the commodity chain are more common than others (see Figure 4.2). For example, around a third (n = 291) of these relationships are with Suppliers and Service Providers, illustrating that most Scottish life science firms are too small to internalize these capabilities and supporting the argument that knowledge 'intermediaries' are vital for such firms (see Cooke 2007). However, the extent of these inter-organizational relationships reveals very little about their geographies.

Second, the location of the external organizations which Scottish life science firms have linkages to illustrates the evolving and multiple geographies of knowledge-based commodity chains. In particular, these results show that there is a need to take a more nuanced view of 'local-boundedness' (see Phelps 2004),

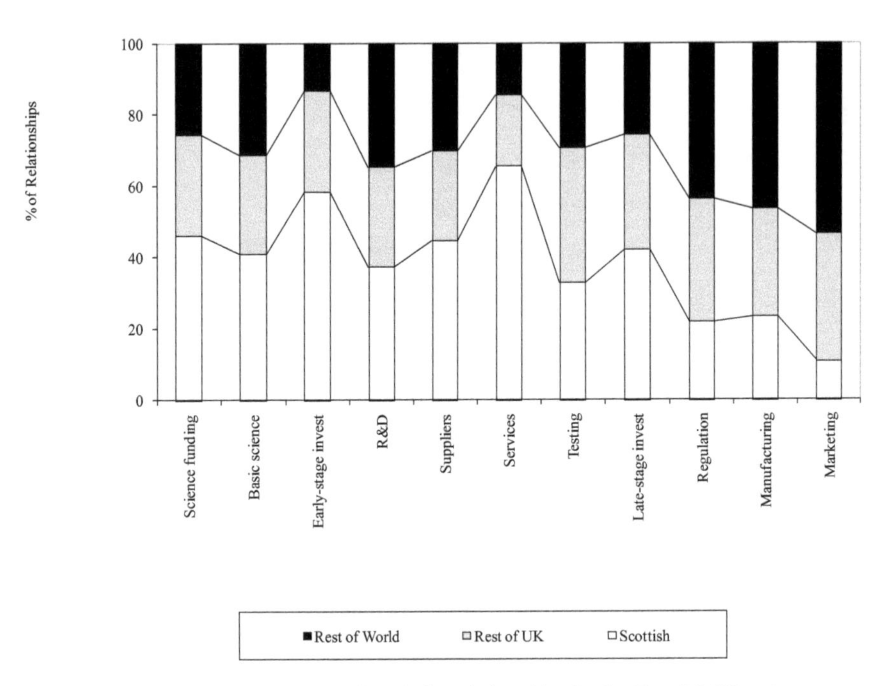

Figure 4.3 Geographies of commodity chain relationships in the Scottish life sciences
Source: Birch (2011), reproduced with permission of Wiley Blackwell.

especially with regards to technoscientific capacity since the surveyed firms had relationships stretching across three broad scales: Scotland (i.e. 'regional'), Rest of the UK (i.e. 'national'), and Rest of the World (see Figure 4.3). The proportion of 'regional' relationships declines further up the commodity chain as firms near the market; in turn, the proportion of international relationships increases. Two other notable findings are worth mentioning: first, most of the science-based relationships (i.e. Science Funding, Basic Science Research, and R&D) are multi-scalar with only a slightly higher regional proportion. This implies that whereas certain knowledge activities can be locally based (e.g. legal advice, accounting services etc.), other knowledge activities (e.g. basic research, R&D etc.) necessitate extra-local linkages that tie firms into wider knowledge networks. Second, only 4 per cent of the firms had an R&D relationship with another Scottish life science firm suggesting that local interaction and learning between cognate firms is limited (see also Moodysson 2008). In contrast, the most common regional relationships were with Suppliers and Service Providers and Early-stage Investors.

Third and related to the last point above, the geographies of these relationships is connected to the type of organization that Scottish life science firms are linked to; a point which was reinforced in the in-depth interviews discussed

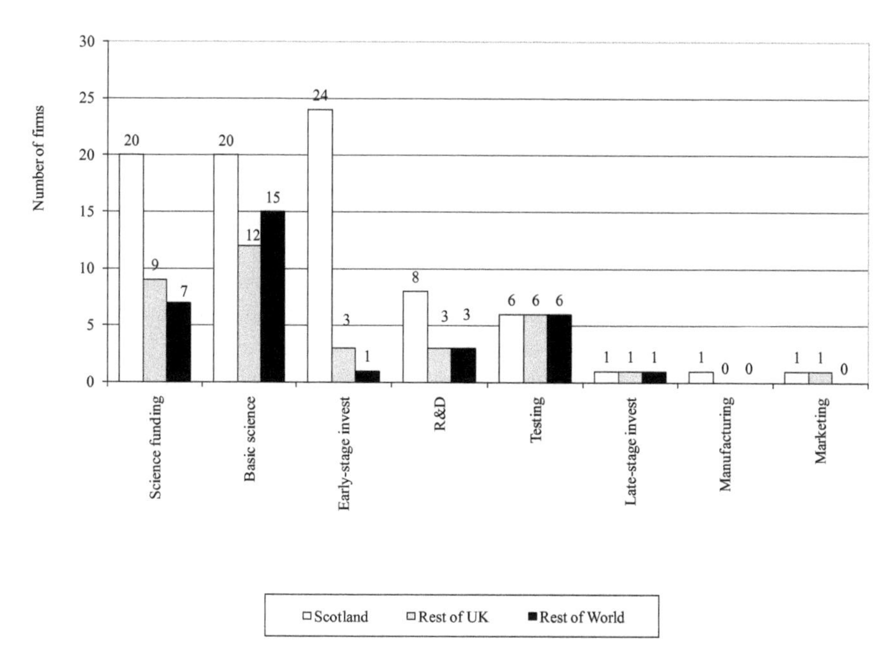

Figure 4.4 Commodity chain relationships to public sector organizations in the Scottish life sciences

Source: Birch (2011), reproduced with permission of Wiley Blackwell.

below. Identifying 'public' organizations revealed an obvious geographical basis to these relationships (see Figure 4.4). First, relationships with public sector organizations are predominantly limited to the early stages of knowledge-based commodity chains; conversely those with private sector organizations are more dominant later in the chain (Birch 2011). Second, the public sector relationships are predominantly regional, whereas private sector relationships are multi-scalar and increasingly global as firms near the market (e.g. Manufacturing and Marketing stages). What this helps to illustrate is the extent to which the public sector (e.g. Scottish Enterprise, Scottish Government etc.) has played an important anchoring role in Scotland, providing financial and other forms of support for both firms and universities. Such early-stage assistance is vital because it enables small firms to initiate new research projects and product development programmes which would otherwise face financing problems as 'traditional' investment sources in the life sciences (e.g. venture capital) are more limited in places like Scotland, especially in comparison to other parts of the UK like East and South-east England (Rosiello and Parris 2009). This finding also illustrates that the institutional context is deeply implicated in the positioning of different parts of knowledge-based commodity chains in different places, a point I will come back to in Chapter 7.

Innovation governance in the Scottish life sciences

Governance, coordination and trust

The complexity of the life sciences as a new technoscientific regime (Coriat *et al.* 2003) is evident in the intricacy and variety of relationships along the commodity chain as outlined above. As noted in Chapter 2 and above, this complexity entails a new form of governance driven by alliance-making and collaborative competencies in which the coordination of specialized and interdisciplinary capabilities and incentives across different organizations is paramount (Birch 2008). At the same time, such competencies and capabilities entail different economic and social relations that are framed by different geographies of governance, coordination and trust. Here, I focus on the regional (i.e. Scottish) dimension of these relations and what this reveals about the agency and potential for local and/or regional actors within the broader knowledge-based commodity chain.

In the life sciences, innovation coordination and governance are shaped by two main factors. First, the complexity of new technoscientific research means that no single organization – whether multi-sited, multi-national or not – is capable of integrating or incorporating all the necessary competencies in their internal organizational arrangements. As a Life Science Alliance (LSA) official put it:

> ... Wyeth [and the Transnational Medical Research Consortium] shows that even the biggest companies in the world need to collaborate, they can't do it all on their own. They used to be able to do it all on their own, but they can't now because the technology is too broad, too many different lines of expertise required, so it really is, you can't do it on your own any more.
>
> (LSA official)

So, as this informant and others pointed out, it is not simply that complexity necessitates collaboration and precludes organizational integration, it is also the concomitant breadth of technoscience and 'different lines of expertise' that drives alliance-making and collaboration across different places. Innovation is a necessarily dispersed process that cannot be concentrated or captured in one place, or by a single set of actors, which requires an active search for complementary knowledge and continual technology upgrading.

Second, there is an evident shift in the form of knowledge production entailing inter-disciplinary, trans-disciplinary and multi-disciplinary working since life sciences firms engage in diverse areas of research. As one Biotech Manager put it:

> ... we don't particularly look to Scotland ... except, that's not true, for our drug discovery, anti-inflammatory drug discovery core we haven't really, particularly looked to Scotland in terms of events or ... we tend to look internationally.

> And on the glycobiology side, that specialist, very specialist area, again it's internationally we look. But, we've entered stem-cell area in terms of R&D and that has been through contacts, not specifically through events, but through [Scottish] university contacts.
>
> (Biotech C)

Knowledge production, for life science firms, entails international interaction and searching across different and sometimes very distinct disciplines (e.g. anti-inflammatory, glycobiology); there is concomitant interaction and searching within more local, Scottish knowledge networks. As a consequence, these Scottish life science firms plug into wider, global knowledge networks which provide them with access to a wider array of expertise and know-how than might be available in Scotland alone.

As the emphasis in the literature on global pipelines (Bathelt *et al.* 2004) has emphasized, such broader spatial connections are critical in the (re)generation of knowledge capabilities. But the main point to emphasize is that it is imperative to retain open and dynamic networks in sectors dependent upon knowledge production. Ultimately the expertise and knowledge necessary for life science innovation cannot be located in any one organization – or even one region – because so many different disciplines, capabilities and assets contribute to the innovation process at the same time that the risk for any one organization of attempting integration is too great. Thus not only will the types and forms of knowledge be different across different organizations and regions, but the processes of learning, upgrading and commercialization will vary at the same time depending on a variety of factors such as – but not limited to – the size, type and reach of each organization.

In this sense, commodity chain coordination is intricately tied to specific geographies of innovation governance in that firms cannot operate alone or *in situ* nor without reference to national and global regulatory regimes and international market demand. Consequently, innovation governance concerns the coordination of socially and spatially dispersed processes within different national and international regulatory regimes. However, this point needs to be more nuanced in order to illustrate the complex relationships that life science firms have with other organizations.

First, life science firms engage in numerous extra-local relationships along the commodity chain in order to access the requisite knowledge and capabilities necessary to move from 'discovery' to 'market', which is especially important for knowledge relations and exchange (see below). As one Biotechnology Manager put it when asked if their research and development partners were based in Scotland:

Biotech B: No, they're based in Germany or France, around Europe.
Interviewer: Is there any particular reason for ...?
Biotech B: Well the righ-, I, I, I wasn't making the decisions in terms of where they were, it will be based on the fact that these guys have the right skill, they can do it at the right cost, they've got the background, the kudos to do it.

Interviewer: And there's been no particular, sort of, issue with them being global or, you know, outside of the UK?

Biotech B: No, not at all. I mean I can tell you, I can tell you that, this may come up later, but I think if you try and run a Scottish life sciences company only using Scottish parts, it would struggle. I've seen that.

This point was reiterated by another Biotechnology Manager who emphasized that such extra-regional relationships do not merely result from the need for international market access, which is centrally implicated in the different geographies of governance, but that there is also a need to connect into broader knowledge networks, as Bathelt *et al.* (2004) have argued and another Biotech Manager claimed:

> Ok, so that's the know-how that it hinges on, but in terms of market, you know, it's a small country, the markets are everywhere else. So, so it's very important to have these connections with the rest of the world. And that's not, not just in, in sort of harsh commercial terms, but in other intellectual terms as well. We have to go out there and make our expertise known globally.
> (Biotech C)

Such networks are necessary because the inherent specialization of life science firms (and collaborating organizations) means that the localized cross-fertilization of ideas is limited, except where there are a number of firms working in *exactly* the same field (e.g. reproductive healthcare); otherwise 'the amount of [local] interaction is limited' (Biotech B). This goes some way to explaining the finding that only 4 per cent of Scottish life science firms had a relationship with a cognate firm. Thus the life sciences require these extra-local linkages from the outset, which would suggest that only those regions with established global 'pipelines' will succeed in stimulating new sectors.

Secondly, small life sciences firms have to tread a fine line between collaborating with other organizations, especially other firms, and maintaining their own distinct and specialized knowledge and expertise. This opens up problems in the arrangement of external relationships since life science firms are specifically reliant upon developing and protecting in-house expertise and knowledge at the same time that they coordinate external collaboration, which is especially important in an industry characterized by the outsourcing of risk (see below). This is evident in the comments of another Biotechnology Manager when talking about the importance of international linkages:

Interviewer: Right, and what do you expect from these, sort of, international events and so on?

Biotech G: More contacts and hopefully interesting collaboration. But I don't think any company will, will explain their technology, I won't get ...

Interviewer: No?

Biotech G: ... ideas from other people. No, that's what technology's about. If you have something new you keep it yourself.

Here the informant points out the need to retain and strongly protect in-house expertise and knowledge, whilst seeking collaborations, often internationally; a comment reiterated by others. It is this in-house capability that is necessary prior to collaboration since, as another informant put it: 'any good biotech has to have a good enough platform of its own that you can then add to through collaborations' (Biotech D). Overall it is evident that there are contradictory pressures on life science firms that drive the establishment of extra-local linkages and the specific form that these take. On the one hand they need to develop and protect knowledge and expertise in-house, whilst, on the other hand, the same knowledge and expertise enables them to connect into broader knowledge and innovation networks along the commodity chain. The governance of these contrasting pressures entails the coordination of different types of relationships and different dimensions of trust discussed next.

Governance of innovation relations and collaboration

Life science firms engage in a variety of relationships along the commodity chain. These linkages, however, cannot be considered as the same as one another because they involve a variety of organizations, different institutional settings, different types of relationship, different geographical dimensions, and so on. All of these are implicated in innovation governance. In this sense then, these linkages entail different geographies of governance in that they involve complex ways of coordinating relationships across different locations.

The governance of knowledge transfer is evident in the emphasis placed on access to global knowledge that was repeatedly referred to by informants, as well as in the 'upscaling' of technoscientific capabilities across organizational relationships, as one Biotechnology Manager pointed out:

Interviewer: So, is this, is this relationship with the Dutch company been fairly important?

Biotech G: It is crucial. Yes. And it is still crucial to maintain the relationship.

Interviewer: So what does it, what does it contribute in terms of the development of the company and the technology?

Biotech G: Because they can, they can upscale the technology, and they can put the technology into manufacturing. They put the technology together with the engineers and then they manufacture a product that they can sell. So they make the technology practical. And so the relationship is still ongoing.

Here, the ability to 'upscale the technology' – and upscaling here is organizational in the first instance though with clear spatial effects – is deemed to be outwith the capabilities of the small life science firm, but not of large manufacturing companies whose size provides economies of scale that make the technology 'practical'; that is, produced at a cost that can be sold for profit. However, it is not simply the size limitations of life science firms – and the associated limitations on certain capabilities that their size implies, as discussed below – that necessitates the shifting of coordination upwards towards large multi-national companies (MNCs) whose size, reach, knowledge and reputation are sufficient to capture the value embodied in the commodity (Gray and Parker, 1998). Knowledge transfer also involves more than simply accessing new knowledge and new capabilities from outside of a firm's 'home' region.

First, the governance of knowledge exchange is underwritten by the *objectification* of trust-based relationships, which contrasts somewhat with the conceptualizations of 'competence' and 'goodwill' trust discussed in the theoretical section above (Cumbers *et al.* 2003; MacKinnon *et al.* 2004). As outlined in the theoretical section, competence trust is based on the idea that trust is performance-based, whilst goodwill trust emphasizes the informal, personal-based nature of certain relationships. That is not to say that these two sorts of trust-based relationships are absent in the Scottish life science; indeed, they are ever present as an Environmental Biotechnology Manager explained:

> And so we have relationships with contractors, and we have preferred suppliers for all that, and we have people we know we can trust, and we have people we have used many times and know that we can use again, and we have ones, by the way, we've used who were a disaster and we'll never use again.
>
> (Environmental Biotech A)

Despite the existence of these forms of trust and the ubiquitous reliance on personal recommendations in the Scottish life sciences, however, there is a strong emphasis on the objectification of firm-level capabilities and activities, which helps to produce a form of *objective trust*. This type of trust contrasts with inter-subjective forms of trust, such as competence and goodwill trust. For example, as one Healthcare Manager, referring to their supplier relationships, put it:

> Well we have to have careful relationships because we're ISO [International Standards Organisation] 9001, so we have to have approved suppliers, and of course, for our regulatory work we need to have properly approved suppliers as well. So we, every single one of ours has to be listed, have an ISO or be inspected or individually audited by us, *so we have good relationships with our suppliers*.
>
> (Healthcare C; our emphasis)

Here the ISO standards represent the basis for the 'good relationships' that the firm has with its suppliers. Notably, it is not the type or form of the relationships themselves that matters, nor is it the closeness of the relationships, nor the previous performance of the firm in question that engender trust. Rather it is the adherence to a particular set of standards embodied in objects and practices that engender trust.

Second, trust is an iterative effect of each firm's need to be regulatory compliant (e.g. to ISO standards) that then drives their need to ensure the regulatory compliance of their suppliers, manufacturers etc.; this is especially so when it involves 'regulatory work', which is work that will ultimately ensure product sales in the long-run. Since product development is such a long process, it is perhaps inevitable that firms build these 'objective' standards into their work from an early stage, which has a recurrent, trust-engendering effect as firms seek 'quality assurance' along the commodity chain as illustrated by a Biotechnology Manager's comments:

> For all of these we have done full Q-A [Quality Assurance] audits, with our, it's another realm of the company who we use for regulatory advice, they also have a kind of, quality assurance division. So they've helped set up our own quality management systems in house, so as part of that supply selection is absolutely key. And because of how important these CROs [clinical research organizations] and our manufacturers are to what we do, we've audited them all, which was interesting.
>
> (Biotech D)

Regulatory standards and expectations represent the key influence underpinning the governance of knowledge transfer as evident in the above comment. So, whilst life science commodity chains are structured by dispersed relations of innovation on the one hand, they are also characterized by multi-scalar regulatory processes on the other. Consequently, since there are a number of different national and supranational regulatory regimes that firms have to operate within, the same firms have to incorporate these different national and increasingly standardized international protocols into their operating procedures at an early stage. This, in turn, engenders trust in their operations, which covers not only tangible product development but also knowledge production as the same interviewee explains:

> When it's in house there are certain ISO requirements for things like data recording. That's one, also how we generate the data. In house we do it to a standard called CLSI, which is the Clinical Laboratory Standards Institute. It's not something that you do have to stick to, but it is internationally recognized, and it's something that will provide a lot of comfort, if you saw data coming across your desk that was done to a particular CLSI method, you should be able to repeat that and get the same data, pretty much under the same conditions. So that's something that we adhere to.
>
> (Biotech D)

The incorporation of these standards into a firm's practices means that other firms can trust the work that the firm is doing. Consequently it is a different kind of trust than competence or goodwill trust, both of which are not based on the objectification of the relationships between organizations.

The issue of multi-scalar regulatory compliance provides a good example of the complexity involved in the governance of knowledge exchange in particular economic geographies, illustrating how the coordination of different capabilities and incentives between organizations is structurally produced and reproduced along the commodity chain. The social processes that constitute this coordination are built around the idea of 'objective trust', which both objectifies certain forms of relationship as trust-worthy – i.e. they are treated as a property of an object rather than social relations – and iteratively informs the practices of life science firms themselves as the 'production' of trust in their own activities necessitates the incorporation of standardized and approved procedures, often set by international organizations, that are accepted in different countries around the world. Consequently, such trust is not constituted by local or even national interaction, linkages and relationships; rather, as a mechanism bound up with the universalizing tendencies in global regulatory regimes, it facilitates the convergence of practices across different geographies. Thus it illustrates the need to go beyond localized analyses in order to understand the complex geographies of economic governance in the life sciences.

Governance of innovation outsourcing

Whereas the governance of knowledge exchange involves the (iterative) incorporation of specific standards into firm-level practices, the governance of risk outsourcing involves both the embedding of distinct, specialized and differentiated knowledge in firm-level expertise and capabilities, and the protection of this knowledge in-house and when collaborating with external partners. Without the protection of in-house expertise there would be no incentive for other organizations to collaborate with a firm since the knowledge would be freely available. Whilst this may sound like it is detrimental to large multi-national corporations (MNCs), the production and protection of knowledge by small life science firms actually has a beneficial effect in that it helps to shift risk back down the commodity chain away from MNCs. This is because product development in high-technology sectors like the life sciences is extremely uncertain and therefore risky for large firms to invest in, especially when they have existing financial pressures from, for example, shareholders concerned with short-term share prices values – a topic I return to later in Chapter 6. One Biotech Manager from a publicly-listed firm pointed out this strategy:

> That's not the model we have, we've licensed and acquired. And, you know, that's what everybody's doing. The amount of organic in-house drug discovery that is generated in your pipeline these days is pretty small, that's why people are buying, spending so much money buying and acquiring.
>
> (Biotech B)

So outsourcing risk involves the coordination of organizational agendas and strategies that, again, entail particular geographies of innovation governance along the commodity chain.

Generally however, the emphasis on in-house expertise is characterized by the development of certain types of business model in which a life science firm is not left dependent upon a single idea or product: one popular hybrid model was to combine product discovery with the provision of services (e.g. screening, laboratory testing etc.). As several Managers pointed out, a strategy based on a single idea or product leaves no room for failure, an all-too-common phenomena in the life sciences (Pisano 2006). Consequently, most small, unlisted firms sought to diversify their capabilities in order to alleviate the risks of failure, although, at the same time and in what might appear a slightly contradictory position, the same firms had largely built up or based their strategy on niche or speciality expertise (see Moodysson 2008). For most firms, especially those working in analytical or science-based fields such as modern biotechnology (Moodysson *et al.* 2008), an inherent feature of this business model was a strategy of out-licensing or partnering since small firms expect to license their expertise and intangible assets to larger firms that would then 'upscale' the technology into products; in turn, large firms increasingly rely upon small firms to provide new knowledge and capabilities that might be risky (i.e. involve opportunity costs) to develop in-house themselves. This is clearly emphasized in the interviews with several Managers:

> *Biotech A*: We want to out-licence as well. I mean we have business model that revolves around early, mid, and long term licensing opportunities.

> *Biotech C*: So, we take molecules from discovery, from isolation, develop them through into clinical trials, basically as far as we can get them, probably to Phase 2a, at which stage we license them to pharmaceutical companies.

> *Medical Devices C*: We are at the point, we are looking, we are looking also for a probably corporate partner, this would be someone who would have the, sort, of, world-wide reach in terms of marketing and product support that we'll never have, we're too small to do that.

> *Medical Devices E*: So, it is basically, it's a research and development skill set to develop products which are then licensed, so actually it is quite a long-term outlook.

Whilst the creation of licensing revenues – the lifeblood of these small life science firms – is thus dependent on the governance of risk outsourcing, in which commodity chain coordination is driven by larger firms as we discussed above, there is a parallel process in which larger firms – especially in the bio-pharmaceutical sector – are increasingly dependent upon the very specialized knowledge outputs and often niche expertise of these small firms. Licensing and partnering arrangements tie these small and large firms together in collaborative and alliance-based relationships shifting governance down the commodity chain through the economic fragmentation of the supply chain.

First, the governance of risk outsourcing is based, to a greater extent than knowledge exchange, on inter-subjective forms of trust such as 'competence trust' and 'goodwill trust' (Cumbers *et al.* 2003; MacKinnon *et al.* 2004).

For example, life science firms commonly relied upon the personal networks of their chairpersons or boards to recruit personnel, to access finance, to make contacts and so on. Although less common, firms would also benefit from other 'non-personal' relationships as a Biotech Manager pointed out:

> Bizarrely we have even been recommended other CROs [Clinical Research Organizations] from their competitors sometimes, if you have a good relationship with them and there's something that they genuinely cannot do, there seems to be a kind of constant circulation of, of people within the industry, and if people have friends that they know can do a good job in another organization, they will actually say, well we can't do this, but I know someone at XYZ who can.
>
> (Biotech D)

Due to the particular out-licensing business model pursued by most life science firms, the relationship between small firms and larger ones was characterized by the coordination of specific incentives which serve to engender trust through the aligning of their different interests. One such example is the use of royalties, rather than outright sale, as part of licences and partnerships, as one Biotech Manager explains:

> And I think it shows a potential partner that yeah if you've got the confidence that you're prepared to focus more on the royalties and the backend of it.
>
> (Biotech H)

Thus, in the light of the uncertain commercial viability of most firms' knowledge assets, the licensing model depends on the *performance* of confidence in one's own knowledge and expertise by accepting specific forms of remuneration; e.g. royalties instead of direct payments. Such coordination of incentives – money and legitimacy for small firms versus new products and expertise for larger firms – illustrates the complex relationships that underpin the governance of risk outsourcing. There is a secondary aspect of this coordination that is particularly beneficial for smaller firms: partnerships, collaborations and licensing deals provide protection from takeovers for small firms since agreements restrict access to their knowledge and capabilities, at the same time that such agreements can build in 'sufficient poison pills', as one informant (Biotech F) put it. However, conversely, the withdrawal of a large firm from an agreement can leave the small firm exposed to a takeover.

Second, such governance is intrinsically tied up with the protection of intellectual and intangible assets; that is, a firm's knowledge and capabilities. Without such protection there would be few opportunities for small firms to collaborate with larger firms and move along the commodity chain. What is striking about this process is the extent to which intellectual property (IP) protection – usually in the form of patents, but also covering internal proprietary *know-how* – represents a 'defensive' strategy as one Biotech Manager explained:

Biotech H: And the patenting strategy is you tend to file base, very broad based patents, and then you begin to file on top of them more specific and exclusive patents, and that's really been our patenting strategy. And, and five of them that we've filed.

Interviewer: Five broad ones or five ...?

Biotech H: Ah five in total, so one broad one and then filing on top of that.

Interviewer: Where did you get that, sort of, the idea for that strategy, or is that just kind of ...?

Biotech H: It's common within the industry. Yep. You're just basically building a wall around what you do so that there's no gaps.

In this sense, inter-subjective trust is thus sanctioned by more formal and contractual arrangements that enable small life science firms to engage in relationships that would be characterized by severe disparities in power otherwise. Another Biotech Manager explained that their firm's R&D strategy was explicitly designed to stop their larger manufacturing partner from finding a loop-hole in their existing IP:

> Very much the R&D's focused on getting more and more protection for our patents, like if the EU appeal fails. So all internal R&D was directed at developing technology so that we can put up new defensive patents. And also cover the patents so that the [partners] do not, maybe, find a loop-hole and manufacture without ... find a loop-hole around our existing patent.
>
> (Biotech G)

Alongside this defensive strategy there is a concurrent need to collaborate with and actually trust other firms and organizations, which entails a clear negotiation over the incentives underpinning the involvement of both sides in any collaboration. In particular, the need for small life science firms to 'own' their knowledge means that they are unlikely to engage with other organizations that do not (or cannot) agree to such concerns. One Biotech Manager put this rather succinctly as follows:

> I think often it's a case of explaining [to the larger firm] why it is that we need to own that [knowledge]. You can use this but we have to own this, and if we can't own that then there isn't any point in moving forward. And generally, you know they do, they do, companies used to working with small companies will understand the need for that and there won't be a problem.
>
> (Biotech H)

The negotiation of distinct incentives and motivations behind knowledge production and, particularly, of IP ownership represents a good example of the complexities involved in the risk outsourcing in that it shows how governance strategies concern both the pursuit of value creation by small firms and the later value capture by larger firms. The social processes that constitute this governance are embedded in inter-subjective forms of trust (e.g. competence, goodwill) that is necessarily underpinned by formal, contractual arrangements (e.g. patents). In this sense governance is not necessarily reliant upon the objectification of social relationships because the underlying governance processes are subject to judgements about trust-worthiness rather than an identification of trust as the property of an object within a framework of regulatory standards. Overall this necessitates the development of closer, personal and direct relationships between individuals, which is less evident in the governance of knowledge exchange discussed above; however, such inter-subjective trust is buttressed by more formal and contractual relationships (e.g. intellectual property rights), which provide a mechanism for engendering trust in relationships that might otherwise be characterized by severe power asymmetries.

Conclusion

My aim in this chapter was to consider whether and how the governance of innovation might offer opportunities for LFRs. In this respect, there is considerable evidence, both in the life sciences and in other high-technology industries, that relations of economic power differ from other less knowledge-intensive sectors because of the need to retain relatively open pipelines – both geographically and organizationally – to new and diverse ideas and knowledge (Bathelt *et al.* 2004). Multi-national corporations find it less easy to dominate and control smaller firms and host regions because the uncertainties around knowledge production tend to work against the concentration and capturing of knowledge. At the same time, the realities of economic power and scale economies remain. Even in knowledge-based sectors, smaller innovative firms need the market reach and global distribution networks of multi-national corporations. Financial capital for firm expansion similarly still tends to locate unevenly and favour the larger and established agglomerations. Hence, the independence offered by the uncertainties of knowledge is always to some extent circumscribed.

On the one hand, life science firms engage in numerous inter-organizational and extra-local relationships because they cannot integrate the necessary knowledge and capabilities in one organization nor will these necessarily be contained in one place. These relationships are constituted by an objectification of trust in that adherence to international and, to a lesser degree, national protocols and standards come to represent 'trustworthiness', which is then iteratively reproduced through the need for regulatory compliance along the whole commodity chain. On the other hand, firms have to ensure that in-house knowledge and expertise cannot be captured by other firms, especially multi-national companies, or they lose their value and thus their ability to engage in collaboration. Such collaborative relationships are highly dependent on interpersonal forms of trust (e.g. competence and goodwill) that enable life science firms to align their interests (e.g. value capture through intellectual property protection) with those of the large multi-nationals that have the necessary global reach to market new technologies. In turn, large companies benefit from this arrangement because they are able to outsource the risk involved in research and development; namely the opportunity costs that long-term R&D necessitates. What this means is that a diverse number of quite different places are able to engage in the KBE, providing nodal points in a broader knowledge network, which supports some of the theoretical arguments made by Bathelt *et al.* (2004) as well as the empirical research of Gertler and Levitte (2005) and others.

References

Allen, J. (2003) *Lost Geographies of Power*, Oxford: Blackwell.

Arora, A. and Merges, R. (2004) Specialized supply firms, property rights and firm boundaries, *Industrial and Corporate Change* 13(3): 451–475.

Asheim, B., Cooke, P. and Martin, R. (eds) (2006) *Clusters and Regional Development: Critical Reflections and Explorations*, London: Routledge.

Bair, J. (2005) Global capitalism and commodity chains: Looking back, going forward, *Competition and Change* 9(2): 153–180.

Bathelt, H., Malmberg, A. and Maskell, P. (2004) Clusters and knowledge: Local buzz, global pipelines and the process of knowledge creation, *Progress in Human Geography* 28(1): 31–56.

Birch, K. (2008) Alliance-driven governance: Applying a global commodity chains approach to the UK biotechnology industry, *Economic Geography* 84(1): 83–103.

Birch, K. (2009) The knowledge–space dynamic in the UK bioeconomy, *Area* 41(3): 273–284.

Birch, K. (2011) ‘Weakness’ as ‘strength’ in the Scottish life sciences: Institutional embedding of knowledge-based commodity chains in a less-favoured region, *Growth and Change* 42(1): 71–96.

Birch, K. and Cumbers, A. (2010) Knowledge, space and economic governance: The implications of knowledge-based commodity chains for less-favoured regions, *Environment and Planning A* 42(11): 2581–2601.

Birch, K. and Mykhnenko, V. (2009) Varieties of neoliberalism? Restructuring in large industrially-dependent regions across Western and Eastern Europe, *Journal of Economic Geography* 9(3): 355–380.

Birch, K., MacKinnon, D. and Cumbers, A. (2010) Old industrial regions in Europe: A comparative assessment of economic performance, *Regional Studies* 44(1): 35–53.

Bunnell, T. and Coe, N. (2001) Spaces and scales of innovation, *Progress in Human Geography* 25(4): 569–589.

Casper, S. and Kettler, H. (2001) National institutional frameworks and the hybridization of entrepreneurial business models: The German and UK biotechnology sectors, *Industry and Innovation* 8(1): 5–30.

Coe, N. and Bunnell, T. (2003) ‘Spatializing’ knowledge communities: Towards a conceptualization of transnational innovation networks, *Global Networks* 3(4): 437–456.

Coenen, L., Moodysson, J. and Asheim, B. (2004) Nodes, networks and proximities: On the knowledge dynamics of the Medicon Valley biotech cluster, *European Planning Studies* 12(7): 1003–1018.

Cooke, P. (2007) *Growth Cultures*, London: Routledge.

Cooke, P. and Morgan, K. (1998) *The Associational Economy: Firms, Regions, and Innovation*, Oxford: Oxford University Press.

Coriat, B., Orsi, F. and Weinstein, O. (2003) Does biotech reflect a new science-based innovation regime?, *Industry and Innovation* 10(3): 231–253.

Cumbers, A., MacKinnon, D. and Chapman, K. (2003) Innovation, collaboration, and learning in regional clusters: A study of SMEs in the Aberdeen oil complex, *Environment and Planning A* 35: 1689–1706.

European Council (2000) *An Agenda of Economic and Social Renewal for Europe* (aka Lisbon Agenda), Brussels: European Council.

Gereffi, G. (1994) The organization of buyer-driven global commodity chains: How U.S. retailers shape overseas production networks, in G. Gereffi and M. Korzeniewicz (eds) *Commodity Chains and Global Capitalism*, Westport, CT: Greenwood Press, pp. 95–122.

Gereffi, G. (1996) Global commodity chains: New forms of coordination and control among nations and firms in international industries, *Competition and Change* 1(4): 427–439.

Gereffi, G., Humphrey, J. and Sturgeon, T. (2005) The governance of global value chains, *Review of International Political Economy* 12: 78–104.

Gertler, M. and Levitte, Y. (2005) Local nodes in global networks: The geography of knowledge flows in biotechnology innovation, *Industry and Innovation* 12(4): 487–507.

Gertler, M. and Vinodrai, T. (2009) Life sciences and regional innovation: One path or many?, *European Planning Studies* 17(2): 235–261.

Grabher, G. (1993) The weakness of strong ties: The lock-in of regional development in the Ruhr area, in G. Grabher (ed.) *The Embedded Firm: On the Socio-Economics of Industrial Networks*, London: Routledge, pp. 255–277.

Gray, M. and Parker, E. (1998) Industrial change and regional development: The case of the US biotechnology and pharmaceutical industries, *Environment and Planning A* 30: 1757–1774.

Haakonsson, S. (2009) The changing governance structures of the global pharmaceutical value chain, *Competition and Change* 13(1): 75–95.

Hopkins, T. and Wallerstein, I. (1986) Commodity chains in the world-economy prior to 1800, *Review – Binghampton* X(1): 157–170.

Hudson, R. (1999) The learning economy, the learning firm and the learning region: A sympathetic critique of the limits to learning, *European Urban and Regional Studies* 6(1): 59–72.

Hudson, R. (2011) From knowledge-based economy to ... knowledge-based economy? Reflections on the changes in the economy and development policies in the North East of England, *Regional Studies* 45(7): 997–1012.

Jessop, B. (2006) State- and regulation-theoretical perspectives on the European Union and the failure of the Lisbon Agenda, *Competition and Change* 10(2): 141–161.

Kitagawa, F. (2004) Universities and regional advantage: Higher education and innovation policies in English regions, *European Planning Studies* 12(6): 833–852.

Lazaric, N. and Lorenz, E. (1998) The learning dynamics of trust, reputation and confidence, in N. Lazaric and E. Lorenz (eds) *Trust and Economic Learning*, Cheltenham: Edward Elgar, pp. 1–20.

Lorenz, E. (1999) Trust, contract and economic competition, *Cambridge Journal of Economics* 23: 301–315.

MacKinnon, D., Chapman, K. and Cumbers, A. (2004) Networking, trust and embeddedness amongst SMEs in the Aberdeen oil complex, *Entrepreneurship and Regional Development* 16: 87–106.

MacKinnon, D., Cumbers, A. and Chapman, K. (2002) Learning, innovation and regional development: A critical appraisal of recent debates, *Progress in Human Geography* 26(3): 293–311.

Malmberg, A. and Power, D. (2005) (How) Do (firms in) clusters create knowledge?, *Industry and Innovation* 12(4): 409–431.

Markusen, A. (1996) Sticky places in slippery space: A typology of industrial districts, *Economic Geography* 72(3): 293–313.

Massey, D. (1995) *Spatial Divisions of Labour* (2nd edition), Basingstoke: Macmillan.

Moodysson, J. (2008) Principles and practices of knowledge creation: On the organization of 'buzz' and 'pipelines' in life science communities, *Economic Geography* 84(4): 449–469.

Moodysson, J., Coenen, L. and Asheim, B. (2008) Explaining spatial patterns of innovation: Analytical and synthetic modes of knowledge creation in the Medicon Valley life-science cluster, *Environment and Planning A* 40: 1040–1056.

Morgan, K. (1997) The learning region: Institutions, innovation and regional renewal, *Regional Studies* 31(5): 491–503.

Ossenbrugge, J. and Zeller, C. (2002) The biotech region of Munich and the spatial organisation of its innovation networks, in L. Schätzl and J. Revilla (eds) *Technological Change and Regional Development in Europe*, Berlin: Physica-Verlag, pp 233–249.

Phelps, N. (2004) Clusters, dispersion and the spaces in between: For an economic geography of the banal, *Urban Studies* 41(5/6): 971–989.

Pisano, G. (2006) Can science be a business? Lessons from biotech, *Harvard Business Review* 84: 114–125.

Powell, W. and Snellman, K. (2004) The knowledge economy, *Annual Review of Sociology* 30: 199–220.

Reich, R. (1991) *The Work of Nations*, New York: Knopf.

Rosiello, A. (2004) *Evaluating Scottish Enterprise's Cluster Policy in Life Sciences: A Descriptive Analysis*, Innogen Working Paper No.16, University of Edinburgh.

Rosiello, A. (2007) The geography of knowledge transfer and innovation in biotechnology: The cases of Scotland, Sweden and Denmark, *European Planning Studies* 15(6): 787–815.

Rosiello, A. and Parris, S. (2009) The patterns of venture capital investment in the UK bio-healthcare sector: The role of proximity, cumulative learning and specialisation, *Venture Capital* 11(3): 185–211.

Senker, J. (2005) *Biotechnology Alliances in the European Pharmaceutical Industry: Past, Present and Future*, SEWPS Paper No. 137, SPRU, University of Sussex.

Storper, M. (1995) The resurgence of regional economies, ten years later: The region as a nexus of untraded interdependencies, *European Urban and Regional Studies* 2: 191–222.

Storper, M. (1997) *The Regional World*, London: Guilford Press.

Warhurst, C. and Thompson, P. (2006) Mapping knowledge in work: Proxies or practices?, *Work, Employment and Society* 20: 787–800.

5 Innovation imaginaries in the European life sciences

Co-authored with Les Levidow and Theo Papaioannou

Introduction

As the previous chapter showed, innovation is a multi-scalar process necessitating specific forms of governance that can coordinate a disparate set of organizational relationships. Focusing on the (promised) commercialization of scientific research in these relationships, however, means that the analysis of innovation and its geographies can end up ignoring equally important issues. One particularly important question that needs addressing when it comes to the geography of innovation is how specific policy visions are enacted as policy strategies and priorities in knowledge-based commodity chains, or any other value chain (Gibbon and Ponte 2008), and at what scale (see Birch *et al.* 2010; Ponte and Birch 2014). As Allison Loconto (2010) argues, this means understanding how visions become 'performative'; in Loconto's case, she focuses on visions of sustainability and their enactment in particular environmental certification schemes or standards. Innovation governance, therefore, entails more than coordinating organizational relations and linkages, it also requires the coordination of various discourses, narratives and visions – which can be conceptualized collectively as *imaginaries* (Jasanoff 2004; Jessop 2005) – implicated in the transformation of societal institutions and policy frameworks. Consequently, understanding imaginaries and their performativity (i.e. how they are enacted) is important for any understanding of the geography of innovation because those imaginaries help to shape global, national *and* regional institutions, policies, and social action.

In this chapter I focus on the emergence of a particular innovation imaginary in the European Union since the mid-2000s; namely, the concept of a *knowledge-based bio-economy*. It emerges from earlier and broader innovation narratives like the 'knowledge-based economy' proposed at the 2000 Lisbon Summit by the European Council (European Council 2000). According to this so-called Lisbon Agenda, new scientific knowledge and technological innovation are the driving forces behind sustainable wealth creation, if only society would adapt to these imperatives (Birch and Mykhnenko 2014). Otherwise Europe will fall behind globally in competitiveness and productivity gains. The Lisbon Agenda was extended to the life sciences, emphasizing the importance of biotechnological processes and products within a distinct institutional and policy framework – that is, the knowledge-based bio-economy (KBBE). According to the EC, the KBBE is 'the sustainable, eco-efficient transformation of renewable biological resources into health, food, energy and other industrial products' (DG Research 2006). Although the KBBE is not a widely known term among European publics, it has been used to define research and innovation policy by linking current problem-diagnoses, research priorities, technological innovation, future visions and societal benefits.

It is important to examine the KBBE as a discursive process because it entails a normative narrative of technological fixes to achieve societal goals and/or to address societal challenges deemed critical for future wellbeing, livelihoods and prosperity (see also Felt *et al.* 2007). The KBBE entails a *performative* narrative that frames these future visions as the rationale for particular institutional and policy changes to achieve this end. In this sense, these visions could end up being self-fulfilling; that is, by creating the conditions for what they seek to promote. Moreover, even though future visions might not necessarily achieve the expected technoscientific outcomes, promotional efforts can nevertheless reshape institutional and policy frameworks.

This chapter draws on concepts from science and technology studies (STS) in order to examine how imaginaries shape innovation policies and institutions. According to STS perspectives, imaginaries are descriptive, normative and performative all at once (see Surel 2000); they describe societal problems, opportunities and challenges in ways favouring specific technological futures. This discursive process naturalizes institutional changes that reinforce a specific innovation narrative. This chapter first outlines these STS perspectives, and then focuses on European policy discourses as the empirical focus for exploring how the KBBE agenda shapes the life sciences sectors.

Imaginaries, technological expectations and self-fulfilling prophecies

I use *imaginaries* as a concept to denote future-oriented narratives, visions and promises. While the concept has a long pedigree, I am more interested in how it has been used recently in the work of people like Jasanoff (2004), Jessop (2005, 2009), Jasanoff and Kim (2009), Birch *et al.* (2010), Fairclough (2010), and Levidow *et al.* (2012). As a concept, imaginaries is useful because it specifically focuses on how imaginings of the future are enrolled in policy strategies and governance in the present. As Fairclough (2010: 266) argues, these future-oriented discourses define what 'could or should be', and the obverse, thereby shaping future innovation goals and plans (Levidow *et al.* 2012). While imaginaries are discursive objects, they are necessarily geographical as well. For example, policy-makers and others draw on geographical imaginaries as well, including 'regional' or 'national' identities, in order to shape future social and economic orders (Jessop 2005; Jasanoff and Kim 2009). Consequently, imaginaries are geographically diverse and varied, in that policy-makers in different places seek to promote different future visions and pursue different policies as a consequence. Conversely, the burgeoning literature on policy transfer (e.g. Peck 2011) demonstrates that certain policy visions and strategies travel from place to place as well.

Innovation imaginaries and technological expectations

For my purposes in this chapter, I apply a number of STS perspectives to innovation imaginaries, drawing from the sociology of (technological) expectations, futures and foresighting (e.g. van Lente 1993; Guice 1999; Rappert 1999; Brown *et al.* 2000; Brown and Michael 2003; Borup *et al.* 2006). In this literature, visions of the future play a constitutive function in enrolling diverse social actors as stakeholders in new technological developments as well as in promoting 'economic imaginaries' more generally (Jessop 2005). Such future visions could be easily dismissed as hype. However, these future narratives, visions and expectations (i.e. imaginaries) are also fundamentally implicated in innovation processes as well as institutional and policy change (Levidow *et al.* 2012).

The role of imaginaries has been analyzed by STS scholars working on the notion of technological expectations. According to Brown and Michael (2003: 4):

> These and other developments emphasize the need for scholarship to engage with the future as an analytical object, and not simply a neutral temporal space into which objective expectations can be projected .... they highlight the need to shift the analytical angle from looking into the future to looking at the future, or how the future is mobilized in real time to marshal resources, coordinate activities and manage uncertainty.

In these STS analyses, future expectations are 'generative' in that they direct activities, provide legitimation and build interest in particular technoscientific solutions to societal problems. Future expectations are crucial in 'bridging' or 'mediating' the boundaries between diverse groups, interests and scales; such expectations help to align diverse incentives and capabilities in pursuit of a common goal (Borup *et al.* 2006). Given the increasing importance assigned to scientific research in knowledge economy agendas, expectations take on a particularly important role in blurring the distinction between societal and economic goals.

Several STS scholars have highlighted the important role of expectations for 'enabling' technologies such as the new life sciences. In particular, such expectations could 'take the shape of generalized promises' rather than simply '*local* promises' (van Lente and Rip 1998: 223), because they are presented as solutions to current societal problems. Such expectations thereby help to direct national, supranational and global research funds towards some innovation trajectories rather than others. Likewise Brown *et al.* (2000) argue that such enabling technologies depend upon enrolling broad publics in innovation, especially because it requires considerable levels of public and private investment. Technological innovations are then celebrated as 'breakthroughs', implying a dramatic solution to an intractable problem, yet with no accountability for the promised expectations:

> For example, the basic dynamics of the futures market means that expectations are capable of generating enormous near-term share value (with which to conduct research or financially reward research staff), but without any necessary requirement for entrepreneurs to fulfil their longer-term promises.
> (Brown and Michael 2003: 13)

More specifically, STS scholars have focused on the importance of expectations in life sciences R&D, where promises and future visions have also been linked to Foucauldian perspectives on the 'economy of hope' (e.g. Helen 2004). The performative nature of expectations is emphasized in several scientific areas: for example, animal biotechnology (Valiverronen 2004); pharmaco- genetics and genomics (Hedgecoe 2003; Hedgecoe and Martin 2003); xenotransplantation (Brown and Michael 2002); and the biosciences more generally (Brown 2003; Brown and Michael 2003). Biotechnology and the new life sciences represent such fertile grounds for these case studies because the sector is plagued by a continuing disjuncture between its 'revolutionary potential' and 'despairing disappointment' (Brown and Michael 2003: 4). Such a disjuncture is clearly illustrated in the work of Nightingale and Martin (2004) who are sceptical about the prospects for new drug development arising from new life sciences.

Imaginaries as self-fulfilling prophecies

According to Brown and Michael (2003: 7), future expectations are often technologically determinist in that 'there is an overwhelming tendency retrospectively to account for success or failure by referring to the properties of a technology or an artefact rather than other equally important factors'. As they further argue, innovation imaginaries mobilize the past:

> ... retrospective memories of the innovation process often forget many of the wide-ranging organizational and material contingencies upon which an artefact's future once depended. Such contingencies are seen as peripheral 'noise' from which the 'victorious' artefactual hero emerges.
> (ibid.)

This determinism represents a major gap in these STS perspectives. The gap concerns the constitutive role that innovation imaginaries play in the emergence or transformation of institutional and policy frameworks. These institutional and policy frameworks are often seen as simple research agendas that present specific futures to be performed:

> Indeed, the emphasis on the need to create a sense of shared future through policy instruments like Foresight programmes are indicative of the need to address and manage the future's fragmentary or indeterminate character. When the future can no longer be expected to follow on neatly from the past, then imaginative means must be employed.
> (Brown *et al.* 2000: 8)

STS literature says little about how institutional and policy frameworks are shaped and driven by particular imaginaries, even when specific expectations are not met. The analysis in this chapter complements the existing literature by exploring this issue in more detail, as well as contributing to an understanding of what such imaginaries mean for the geography of innovation.

It is important, in this regard, to highlight the performative role of self-fulfilling prophecies, especially in the way that institutional and policy frameworks are shaped by technological (and economic) expectations (Surel 2000; Smith 2005). This builds on the distinction made by Borup *et al.* (2006) between two conceptions of expectations: 'realist' and 'performative' (or 'constitutive'). The former presumes that expectations have either a true or fictitious basis. This realist account, and perhaps the concept itself, originated with Robert Merton:

> The self-fulfilling prophecy is, in the beginning, a *false* definition of the situation evoking a new behaviour which makes the original false conception come 'true'. This specious validity of the self-fulfilling prophecy perpetuates a reign of error. For the prophet will cite the actual course of events as proof that he was right from the very beginning.
>
> (Merton 1968: 477)

Such a realist account is problematic when applied to imaginaries (e.g. promises, expectations, visions), whose true or false basis – i.e. what innovation will happen in the future – cannot be ascertained in advance (Borup *et al.* 2006). Innovation can only be made feasible and ultimately real by performing specific imaginaries, to some extent in advance.

Along the latter lines, the performative notion acknowledges the indeterminate, contested nature of future expectations, visions, promises, etc. It is through performative contests that 'science and technology is [sic] actively created' (Brown *et al.* 2000: 3; also Brown 2003: 17). So a performative concept of self-fulfilling prophecies is analytically more useful than a realist account. Moreover, a performative role can reshape more than innovation processes. When new technologies fail to deliver on their economic promises, failure has been interpreted as a sign of insufficient effort in implementing the original policies and institutional changes, rather than as grounds to question those prescriptions. Thus failure can reinforce particular institutional and policy prescriptions (cf. Birch 2007). This performative role potentially reshapes institutional and policy frameworks, which can be defined as follows:

> ... the wider social, political and economic geographies [or context] in which decision-makers work. Such policy work consists of policy analysis, the construction of policy visions, the outlining of policy prescriptions, the modes of policy implementation, amongst other things.
>
> (Birch *et al.* 2010: 2901)

For example, EU policy elaborates a master narrative centred on the 'economics of technoscientific promise'. This helps to attract resources – financial, human, political, etc. – by conflating innovation with societal progress. In this way, EU policy narratives seek to 'justify interventions and pre-empt disruptive public responses' (Felt *et al.* 2007: 22, 75–76). That performative role is evident as well by the KBBE as an agenda for the agri-food sector. EU policy frameworks have been reshaped as a supposed imperative for addressing societal problems through the life sciences and its genetic-molecular reductionist solutions (Birch *et al.* 2010). At the same time, significant dissent from civil society gained policy support for agro-ecological alternatives, thus accommodating divergent accounts of the KBBE (Levidow *et al.* 2012).

Innovation imaginaries in the European life sciences

Methodological note

Building on the STS perspectives discussed above, in the rest of this chapter I analyze EU-level research and innovation policy for the biopharmaceuticals sector, set in the wider context of the KBBE agenda. The empirical material underpinning the analysis comes from policy documents produced by the EU, EC and other European policy actors, especially from European Technology Platforms (ETPs), their predecessors and supporters. The European Council (2003: 14) encouraged the creation of such platforms as a means to bring together 'technological know-how, industry, regulators, and financial institutions to develop a strategic vision for leading technologies'. ETPs were meant to involve 'all relevant stakeholders' as a basis for the identification of societal needs and benefits (ibid.). In turn, these platforms formulate 'common visions' of a future Europe, as the basis for 'Strategic Research Agendas' which can be incorporated into Framework Programme 7 funding priorities. As a result, ETPs can be seen as shaping research and innovation priorities through specific narratives of the future (i.e. imaginaries).

In order to process the data gathered from ETPs, I focused on core categories emphasized in the STS literature: technological promises; futures; expectations; and visions. Various coding techniques were employed to label conceptual categories and properties and unveil logical connections; identify, classify and describe policies; distinguish between core and non-core categories and policies which helped to synthesize ideas and provide answers to research questions. Before I turn to the analysis of the KBBE agenda, however, I outline the broader context in which it has been promoted.

EU research and innovation policy

The EU has a longstanding role in promoting research across its member states in response to the perceived technology or innovation gap with leading industrial nations like the USA and Japan (Sanz-Menéndez and Borrás 2000). Concern with this perceived gap stretches back to at least the 1970s when the Directorate-General for Research, Science and Education was established, although it is also evident earlier in areas like nuclear energy (Georghiou 2001). One particular aspect of this early EU research policy was the aim to encourage collaboration across member states in an attempt to create economies of scale, although this sometimes conflicted with the focus on national champions (Sanz-Menéndez and Borrás 2000). What became evident as EU research policy evolved was the extent to which it was tied to the broader single market integration project, especially with the establishment of the Framework Programme in the 1980s. This led to greater emphasis on innovation as 'research' policy was reconstituted as 'innovation' policy in the 1990s, leading to a greater focus of research and innovation policy on private sector, industrial interests and, especially, the global competitiveness of European companies (Sanz-Menéndez and Borrás 2000; Georghiou 2001). As Susana Borrás (2003) argues, this has meant that global competitiveness has come to underpin EU research and innovation policy, with a particular concern about 'lagging' behind other countries driving policy-making.

As a result of its research and innovation policies, the EC has increasingly emphasized new knowledge-based sectors such as the life sciences. This emphasis fits into the wider concern with the threat of international competition brought about by economic globalization mentioned in Chapter 1, which in turn is used to justify a wholesale reorientation of European institutional and policy frameworks. Since at least the early 1980s, EU policy agendas and visions have presented new technologies as the driver of global competitiveness (Gottweis 1998). According to the EC's 1993 *Growth, Competitiveness, Employment* White Paper, for example, future European competitiveness depends on the 'knowledge-based economy' (KBE), especially high-tech sectors like 'biotechnology'. This represents one of three key industrial sectors 'offering the greatest potential for innovation and a particularly rich source of growth' (CEC 1993: 89). Drawing an analogy with information technology and its systemic productivity impacts, this Commission report assumes that biotechnology has beneficent properties: 'Comparable changes in productivity will be achieved by further progress in life sciences – biotechnology – through the creation of innovation in highly competitive areas of industry and agriculture' (ibid.: 92).

At the 2000 Lisbon meeting of the European Council, Ministers committed the EU to become 'the most competitive and dynamic, knowledge-based economy in the world, capable of sustainable growth with more and better jobs' within a decade. Known as the Lisbon Agenda, this emphasized that 'technology and research represents tomorrow's jobs' (European Council 2000: 18). Two years later, at the 2002 Barcelona European Council meeting, Ministers further emphasized the importance of 'frontier technologies' as a key factor for future growth. In a mid-term assessment of the Lisbon Agenda, the 2006 *Aho Report* emphasized that Europe faces the threat of 'losing out as large firms globalize their R&D'. As a remedy, the report argued, Europe must 'provide an innovation-friendly market for its businesses'. One proposal in the *Aho Report* was that Europe could gain a competitive advantage in 'lead markets'; these would be facilitated by several institutional and policy changes, especially 'a cultural shift which celebrates innovation'. This diagnosis carried a threat and opportunity: 'Europe and its citizens should realise that their way of life is under threat but also that the path to prosperity through research and innovation is open if large scale action is taken now by their leaders before it is too late' (CEC 2006: 2, 6, 25).

These kinds of concern are evident again in the EC's *Life Sciences Strategy* which envisaged modern biotechnology as the 'next wave of the KBE' (CEC 2002: 3). As an enabling technology it would contribute significantly to the success of the Lisbon Agenda. For example, by the 2007 mid-term review of this life sciences strategy, biotechnology became a key instrument for solving diverse 'challenges' covering 'health, energy supplies, global warming and an ageing population' (CEC 2007a: 2). This life science strategy was extended in 2005 when the EC held a high-profile conference launching a research agenda for the *Knowledge-Based Bio-Economy* (KBBE). This KBBE concept combined the EC's earlier KBE agenda with the OECD's proposal for a *bio-economy* 'policy agenda' (OECD 2005). The EC's new KBBE vision frames the life sciences as stimulating, linking and capturing new global markets, thus going beyond the simple introduction of new products. For example, according to the EU Science and Research Commissioner, the life sciences 'is a sector estimated to be worth more than €1.5 trillion per year', involving 'different sciences and technologies, different industries, and different policy areas' (DG Research 2005: 1).

As an EU-level research and innovation policy vision, the KBBE promotes wider policy and institutional changes. According to its diagnosis, societal and policy obstacles must be removed in order to exploit new resources through technoscientific innovation (CEC 2005). The KBBE discursively naturalizes such innovation as necessary for the economic competitiveness on which European prosperity depends.

The KBBE agenda as an innovation imaginary

In analyzing the KBBE as an innovation imaginary, this chapter identifies diverse framings, expectations, and visions underpinning the calls for institutional and policy change. The analysis looks at how this discursive combination makes the KBBE novel and distinctive. Like many other future visions analyzed in STS literature (Brown and Michael 2003; Borup *et al.* 2006), the KBBE agenda features three main elements: *descriptive accounts* of objective reality as threats and opportunities; *normative accounts* of necessary or desirable responses to that reality; and, *performative accounts* for carrying out those responses (see Surel 2000). I analyze the KBBE agenda from an STS perspective as it relates to biopharmaceutical research and innovation policy in order to illustrate how these three elements combine to shape institutional and policy frameworks. Core elements are outlined in Table 5.1.

As originally launched, biopharmaceutical innovation was included as a key part of the KBBE agenda (DG Research 2005). The biopharmaceutical Technology Platform eventually became the Innovative Medicines Initiative (IMI). The IMI was originally launched as a technology platform in May 2004. It aimed 'to reinvigorate the European bio-pharmaceutical sector and to make Europe more attractive for private R&D investment in this sector, by overcoming the research

Table 5.1 The European Commission's KBBE agenda

	Characteristics
Economic lag: indicators and benchmarks	New products: Few have been commercialized. R&D investment: below Lisbon objective and 'lagging' behind USA Patents: also lagging.
R&D deficiencies as diagnosis of lag	Public and private sectors have inadequate links for turning knowledge into products. Inadequate knowledge results in 'systemic' failure in R&D process. R&D expenditure has a rising ratio to new products, so they become more expensive.
Blockages and policy change	Need 'a market for innovation' = for new drugs. Potential products have high failure/attrition rates, thus increasing opportunity costs. Therefore need better methods to reduce attrition. Change policies for reimbursement (e.g. public-sector purchasing).
Natural resources: scarce or abundant	Genomics data can generate efficacious and safe products – e.g. through personalized medicine, tailor-made drugs, etc. – promoting health. Need to validate new techniques for predicting efficacy and safety of new drugs, by bringing together generic knowledge from across the sector. Medical treatments depend on such knowledge.
Societal benefits	Biomedical research will amplify the demographic problems of ageing societies, but personalized medicine will improve quality of healthcare. New medical treatments will provide both economic wealth through employment and better health for patients. *[Disease prevention and public health are relegated to diagnostics.]*

Source: Birch *et al.* (2014); reproduced with the permission of Springer Link.

bottlenecks that currently hamper the drug development process' (DG Research 2007: 51). Unlike most Technology Platforms, the IMI was subsequently established as a Joint Technology Initiative (JTI), representing a public–private partnership between the European Commission and the European Federation of Pharmaceutical Industries and Associations (EFPIA). The IMI's agenda has been steered by the EFPIA from its original position paper (EFPIA 2004a) through to the Strategic Research Agenda (IMI 2008). Although the EFPIA claims to promote 'European competitiveness', one-third of its 39 full company members are US-based multi-nationals, which press for similar institutional and policy changes on the other side of the Atlantic.

For the Framework Programme 7 period (2008–2013), the IMI represented a €2 billion investment – half coming from the EC and the other half provided 'in kind' by EFPIA member companies. The IMI seeks to reduce the 'attrition rate' of biopharmaceutical products in the later stages of development by developing 'pre-competitive' research rather than specific treatments or products:

'These collaborations conduct pre-competitive research projects, i.e. research where companies are not averse to their competitors having equal access to the results' (IMI 2008: 8). Thus there is an emphasis on improving the 'properties' of biopharmaceutical drugs through efficacy and safety evaluations in drug development, as well as improving knowledge management and education and training throughout the drug development process.

Descriptive accounts: 'Lagging behind' diagnoses

When it comes to biopharmaceuticals, the KBBE agenda reflects the concerns of earlier EU research and innovation policy about Europe lagging behind other economic powers, especially when it comes to global competitiveness (Borrás 2003). In particular, descriptive accounts of the problems facing the European biopharmaceutical sector frame the problems and opportunities in specific ways.

First, the KBBE agenda frames the European biopharmaceutical industry as lagging behind the USA in terms of the relationship between 'market' failure – especially the unwillingness of pharmaceutical companies to invest in basic research – and falling R&D spending in Europe. Here the KBBE agenda blames market conditions and regulations, especially difficulties with patenting and new drug pricing, for industry decisions to locate new R&D investment outside of Europe. In particular, the IMI (2004a: 7) position paper claims that the 'slow uptake of medicines' has negative implications for European innovation levels because it 'impacts on the profitability of the pharmaceutical industry'. What is interesting about this claim is that it ignores the global nature of major pharmaceutical multinationals in that they do not need to develop drugs in a particular country in order to sell in that country. Furthermore, the account of falling R&D investment in Europe is also not supported by data that the IMI itself relies on to make its argument. In its Strategic Research Agenda (SRA), the IMI (2008: 21) claims that there is a 'relative under-investment' in European biomedical research. However, a graph shows that European pharmaceutical R&D spending rose from €17.8 billion in 2000 to €21.7 billion in 2005, compared with a rise from €23.1 billion to €25.3 billion in the USA in the same years. In light of this graph, the IMI claim there seems to be some tension with the claim that there is 'relative under-investment' in Europe.

Second, the KBBE agenda frames the problem of drug attrition rates and low product approvals as the result of 'systemic' failures in the R&D process, especially upstream from product approval. Technoscientific knowledge, in the form of 'pre-competitive research', is presented as the solution to 'systemic' problems in the drug development process. In particular:

> The most common factors resulting in project failure have been reported as either a lack of efficacy (25%), clinical safety concerns (12%) and toxicological findings in pre-clinical evaluation (20%) ... The biopharmaceutical industry's greatest need is for failure to be predicted at the earliest possible stage of the drug development process. Advances in basic biomedical science within

> the entire European research community could, potentially, make a significant contribution to improving the predictability of the biomedical R&D process. The vision for the future would be to possess the ability to identify lack of efficacy as soon as possible, even when a drug has promising pre-clinical data, and the potential for adverse drug reactions and pre-clinical toxicity.
>
> (IMI 2008: 10)

There is another tension in this second descriptive framing. This becomes evident in the early claim that the '[i]nnovative capacity of [the] pharmaceutical industry is largely determined by a specific external environment in which the industry operates' (EFPIA 2004a: 2) and the later identification of policy and regulation as obstacles, causing 'potential disturbances in the generation of needed innovation in medicines' (ibid.: 2). This reasoning implicitly equates innovation with particular products resulting from innovation-enhancing policy and institutional arrangements, and a lack of products with innovation-obstructing policies and institutions. Technological developments are deterministically assumed to follow a particular pathway that can be enhanced or obstructed – supporting the findings of Brown and Michael (2003) – but which does not result from social, economic and/or political drivers. According to this account then, the main remedy is to change institutional and policy frameworks so that innovation can successfully create new knowledge for new treatments by reducing attrition rates. For instance, the IMI advocates an IP policy that makes it easier for IMI participants to licence ownership rights in both 'Background' and 'Foreground' knowledge, as if this inherently promotes knowledge creation (IMI 2007).

A third descriptive framing is also evident in the IMI narrative. It veers into a normative account as well (see below) by framing 'health' with 'wealth'. Here new treatments are assumed to ensure the public's health at the same time that they ensure economic growth through technological developments across Europe. For example Janez Potočnik, then EU Commissioner for Science and Research, welcomed the IMI as follows:

> I'm happy to see that this unique public–private partnership that was launched as a new instrument for research two years ago is bearing fruit. In times of crisis, such a model of cooperation is proving particularly well suited to answering both EU public health and economic needs.
>
> (EFPIA 2009)

In the KBBE agenda, future health and wealth are threatened by the economic lag between Europe and other countries, especially the USA, as well as the failures in the innovation process (see above). An early EFPIA position paper identifies the key barrier to innovation as 'the slow uptake of new medicines which impacts the profitability of the pharmaceutical industry'; therefore the 'ultimate goal needs to be to increase uptake of new medicines and return on investment for the industry in Europe' (EFPIA 2004a: 7, 9). Likewise the draft IMI Strategic Research Agenda claims that Europe has lost its 'leading place' as a site for biomedical research

(EFPIA 2005). Consequently the IMI is cast as contributing to the broader Lisbon Agenda, but only if the science base is strengthened. This means 'revitalising' biopharmaceutical R&D in Europe, so that it can 'become competitive' (ibid.). This depends on the transformation of institutional and policy frameworks – e.g. towards a single European patent, free (and therefore higher) pricing, pharmaceutical regulation, etc. – in order to create more attractive market conditions (IMI 2006). These claims are also repeated in the 'final' research agenda (IMI 2008).

In this descriptive account, the European biopharmaceutical industry is falling or 'lagging' behind, as repeated in reviews of the IMI, leading to a loss of economic competitiveness and, as a consequence, of health benefits. By contrast to the US biopharmaceutical sector, this account represents the European sector as suffering from systemic and market failures that the IMI will solve. Ultimately, the agenda conflates systemic failure (i.e. high drug attrition rates and low product approvals in Europe) and market failure (i.e. European companies' unwillingness to invest in R&D) with the fragmentation of both Europe's market and its research and innovation policy.

Normative accounts: Necessary societal responses to research bottlenecks

Following the descriptive problem-diagnoses above, the IMI has the main objective of 'reinvigorating' European research and development by overcoming specific 'bottlenecks' in development technologies that contribute to reducing attrition rates in later R&D stages. These difficulties are highlighted in both the EFPIA's 'vision document' (EFPIA 2004b) and the EC's Third Status report on the Technology Platforms (DG Research 2007). However, nearly 30 per cent of the high attrition rates result from 'commercial reasons' and 'cost of goods' (EFPIA 2004a) – problems which are nowhere discussed but become more evident when specific institutional and policy changes are promoted (see below). The proposed societal responses, however, focus more on the impact of lagging investment and innovation on Europe's economy. For example the second-draft strategic research agenda warns that action must be taken now 'before it is too late' (EFPIA & IMI 2006: 22; cf. CEC 2006: 3, 25). Such threats also offer opportunities through solutions, as the same document argues: 'Examples of areas of science and technology where Europe needs to invest today so as not to face a gap analogous to that what we see in ICT include biotechnologies, cognitive and neuro-sciences' (ibid.).

The KBBE agenda promotes societal responses as illustrated by the 2007 Cologne Summit of the German Presidency which declared, 'Europe has to take the right measures now and to allocate the appropriate resources to catch up and take a leading position in the race to the Knowledge-Based Bio-Economy' (EU Presidency 2007: 6). This resonates with earlier concerns in EU research and innovation policy (Georghiou 2001; Borrás 2003), and reflects earlier EU concerns with global competitiveness. The normative account thereby justifies a re-orientation of institutional and policy frameworks towards technoscientific developments that are underpinned by the objective of global competitiveness (see also Smith 2005). This aim is based on the expectation that innovative bio-based products will resolve the societal problem of European market competitiveness, thereby increasing economic growth and general prosperity. As such, it reflects the arguments made by Brown and Michael (2003) about technological expectations, but, in contrast to much of this STS literature, such expectations are largely economic rather than technological. For example, according to the revised *Action Plan for the Life Sciences*, EC policies need to be 'refocused in order to foster market development for bio-based products and improve the uptake of new technologies' (CEC 2007a: 8). A position reiterated in the claims made by the IMI (see above).

Transforming Europe's institutional and policy frameworks is meant to produce societal benefits by promoting better healthcare, which is conflated with more R&D investment and the sale of biopharmaceutical treatments. This rather linear narrative links public health with market wealth (e.g. CEC 2007c). For example:

> The aim of IMI is to support the faster discovery and development of better medicines for patients and enhance Europe's competitiveness by ensuring that its biopharmaceutical sector remains a dynamic high-technology sector.
>
> (IMI 2008: 3)

In a similar claim, health is meant to improve as a consequence of new medicines, especially where these can be accessed more rapidly through faster development and approval processes (Gillespie *et al.* 2007). To facilitate faster product approval, the KBBE agenda promotes the transformation of regulatory institutions and norms, especially through closer relations between public and private sectors. Here it is evident that the narrative is about institutional change and not necessarily about expected technological developments, as emphasized in the STS literature (Borup *et al.* 2006). Thus the IMI aims to:

> Develop a partnership with regulators to devise innovative clinical trial designs and analyses, to aid acceptance of biomarkers and to promote data sharing and the joint consideration of ethical issues.
>
> (IMI 2008: 10)

These normative accounts also suggest that societal benefits from new medicines will result from the sharing of particular kinds of pre-competitive knowledge (e.g. genetics, genomics, etc.). For example, according to a Commission review of the IMI, 'Dissemination and sharing of results is increasing the innovative value of research and maximising the social return' (CEC 2007c: 28). These expectations are, like those discussed in other studies (e.g. Brown and Michael 2003), technologically deterministic in that they assume new drugs and treatments will improve the quality of healthcare through specific interventions such as personalized medicine (see EU Presidency 2007: 9). This framing largely ignores disease prevention or public health, except where they favour development of new diagnostic products, thereby aligning societal benefits with a particular techno-fix. The KBBE agenda has been criticized on such grounds elsewhere (e.g. Birch *et al.* 2010).

According to an NGO report, the focus on genetic knowledge has produced screening tests which are 'poorly predictive of common diseases and most adverse drug reactions'. As more fundamental criticisms:

- 'Pre-competitive' subsidy, via research funding decisions, lacks accountability and transparency and hides political and commercial commitments to the bio-economy and to imaginary markets presumed to be created in the future;
- Public–private partnerships and public procurement policies shift investment risks and externalities on to the taxpayer, intermediaries such as farmers, doctors and health services, and members of the public (Genewatch UK 2010: 9).

Such criticisms have a long history but have remained marginal to policy debate and have failed to inform policy in the biopharmaceutical sector.

Performative accounts: Institutional and policy changes sought

The KBBE agenda promotes the idea that industry needs better technoscientific capacities and that this can only be achieved by changing institutional and policy frameworks; this contributes to the STS literature by going beyond the technological focus of many studies (Brown and Michael 2003; Borup *et al.* 2006). For example, pre-competitive research is meant to solve systemic failures prevalent in drug discovery and development. In particular, the identification of 'key bottlenecks' is associated with 'pre-competitive barriers to innovation' (EFPIA 2004b: 5); barriers include safety, efficacy, knowledge management and training/education. In this context, failure means high attrition rates and the high costs of new drugs (EFPIA 2005). High drug costs are characterized as an issue only in terms of governments' unwillingness to pay high prices for new drugs, meaning that this is presented as another kind of bottleneck to sit alongside the more technical ones. The solution, therefore, is to change policies and/or institutions rather than the high price of drugs.

As repeated in later documents, such diagnoses form the basis of the final IMI research agenda (e.g. EFPIA 2005; DG Research 2007; IMI 2008). The emphasis on industry's interests (i.e. profit) is evident in the analysis of Europe's lagging performance. This 'innovation gap' is discursively framed around the 'opportunity costs' of drug development, especially in pre-competitive research, which contribute to rising attrition rates. Thus the external review of the IMI states that:

> Working on the problems of attrition, IMI is about more efficient processes for creating drugs (decreasing opportunity costs). Even though it does not necessarily bring cheaper or more drugs, it will develop leads, so that more products can make it to the market place.
>
> (Gillespie *et al.* 2007: 36)

The deficiencies in drug development technologies are thereby attributed to greater opportunity costs (i.e. an economic issue and not a technological one), whose reduction should lead to higher R&D productivity – by contrast to the earlier concerns with increasing European R&D investment and the failure of development technologies themselves. According to the discursive framing, public–private collaboration can lower attrition rates and thus allow Europe to catch up:

> This would enable Europe to become a preferred location for biopharmaceutical industry investment. In recognition of this opportunity, the Research Directors Group of EFPIA has identified the pre-competitive barriers in the drug discovery and development process on which the biopharmaceutical industry, as well as other stakeholders in the biomedical R&D process such as academia, can collaborate.
>
> (IMI 2008: 9; also Ragan 2007)

Consequently, for improving biopharmaceutical performance, the main solution to systemic failure is 'pre-competitive' research featuring new partnerships between public and private R&D (DG Research 2007; IMI 2008). This diagnosis avoids the need to examine competitive research (e.g. the blockbuster model) or market failure (e.g. cost of basic research) or the high cost of new biopharmaceutical products; consequently institutions and policies are expected to be problematic and therefore in need of change now. As the early EFPIA position paper lamented, the 'willingness to pay a fair price for new medicines, particularly in Europe, is rapidly decreasing', creating a major systemic challenge (EFPIA, 2004a: 2).

These diagnoses are performative in that they aim to shift the institutional and policy frameworks that underpin biopharmaceutical development. For validating new medicines, a priority is new 'endpoints' and new technoscientific tools, together enabling regulators to assess (and approve) novel treatments more easily and more positively (ibid.). By promoting new institutional and policy frameworks like IMI, the KBBE agenda supports research that can create these 'pre-competitive' tools through a new drug discovery and development paradigm (e.g. EFPIA & IMI 2006). This means enrolling new participants (e.g. public and private actors) into 'pre-competitive' R&D that will enable the validation of new technoscientific treatments, facilitating commercial products and their approval. In the early plan, 'The identified gaps and bottlenecks will be addressed by new technologies and new paradigms for assessment of safety and efficacy as well as for medical practice' (EFPIA 2004b: 19) – this helps to align the interests of regulators with those of industry. Likewise the IMI is expected to identify 'relevant bottlenecks in the drug discovery and development value chain' (IMI 2008: 34). In particular it will:

> ... deliver powerful new multi-disciplinary tools to improve the innovation process and thus establish a new drug development paradigm ... The potential of IMI to create a new drug discovery and development paradigm is based on a more systematic use of biomarkers and on applying innovative technologies such as omics technologies and other types of data.
>
> (Ibid.: 13, 15)

Such new tools and 'data' are fetishized as the key resource to achieve techno-scientific solutions that rework existing institutional and policy frameworks such as regulatory agencies, university research funding, and so on. In particular the need to enrol European regulators as stakeholders in determining these new tools and standards is presented as a rational response to innovation failures in the biomedical industries. A focus on regulatory obstacles easily justifies more lax cost-effectiveness criteria for the public sector to fund new drugs (i.e. pay higher drug prices) and easier access to patents, as in the UK government's proposals involving biopharmaceutical companies (OLS 2009).

Within the 'market failure' diagnosis, no single firm or combination of firms will pursue such knowledge without encouragement and support from governments, hence why the IMI must necessarily enrol and support public *and* private research capacities. However, another remedy is industry's active participation in creating new institutional and policy frameworks to ensure that the technoscientific solutions to Europe's problems are successful. One such example, as formulated by the then Directorate-General for Enterprise, is for the creation of 'lead market initiatives' that give priority to specific technological sectors such as the life sciences. According to the Commission, competitiveness problem results from inadequate product development and industrial take-up, in turn resulting from 'uncertainty about product properties and weak market transparency'. That diagnosis warrants 'incentives for the emergence of the bio-based product market', e.g. by assessing and possibly modifying EC regulations 'in order to increase efficiency'. It also mandates public–private partnerships in which investors and other stakeholders could participate (CEC 2007b: 2, 9). Industry lobbyists, such as the European Association for Bioindustries (EuropaBio), advocate such policy changes as means to stimulate markets:

> By developing support policies and measures that will stimulate the demand for these products, this new policy ['lead market initiative'] will encourage innovation in bio-based products by transforming knowledge in new bio-products and bioprocesses.
>
> (EuropaBio 2008)

This analysis goes beyond STS perspectives that see technological expectations as neglecting 'equally important factors' beyond technology (Brown and Michael 2003: 7). Ostensibly about research priorities, the KBBE agenda also warns against lost opportunities by referring to factors beyond technology, which thereby can be absolved of blame for any failure. Blockages are identified in current policy frameworks and institutional arrangements – which therefore must be changed along lines more favourable for economic competitiveness.

Conclusion

In analyzing the KBBE in this chapter, I explored several related questions about dominant policy imaginaries – how they define societal problems in ways favouring preferred futures, how they promote particular institutional and policy frameworks, and how they involve self-fulfilling prophecies. Together with the European Technology Platforms as partners, the European Commission presents ‘common visions’ for specific research agendas as a means to address ‘societal needs and benefits’. As one such example, the KBBE agenda creates a common identity, both within and among bio-research institutions, thus informing ‘communities of promise’ (cf. van Lente 1993). The KBBE agenda frames socially relevant bio-knowledge in terms of pre-competitive research which can provide generic knowledge, eventually facilitating patents and new commercial products. The KBBE agenda defines problems (e.g. global competitiveness), recognizes threats (e.g. emerging economies), and identifies opportunities (e.g. renewable resources) along lines that justify its focus on molecular-level knowledge. For example, ill health is attributed to inadequate productivity, efficiency and innovation in biomedical research and development.

In extending STS perspectives, I analyzed how the KBBE agenda emphasizes obstacles outside of research and innovation themselves. The KBBE identifies policy frameworks and institutional arrangements that must be changed to facilitate technoscientific solutions and thus expected (and promised) societal benefits. Through circular reasoning, such agendas can justify further changes along similar lines. Perversely, even technological failure can vindicate the agenda, by justifying greater efforts to overcome supposed policy obstacles. Thus technoscientific expectations, such as the KBBE, can serve as a self-fulfilling prophecy, regardless of whether or not a particular innovation pathway yields commercial products and social benefits. As a narrative of a better future, the KBBE plays a role in justifying specific policies, whilst pre-empting alternative futures and critical responses. Research and innovation policy warrants critical scrutiny in order to understand how such imaginaries shape institutional and policy frameworks, reinforcing particular political-economic assumptions.

References

Birch, K. (2007) The virtual bioeconomy: The ‘failure’ of performativity and the implications for bioeconomics, *Distinktion: Scandinavian Journal of Social Theory* 14: 83–99.

Birch, K. and Mykhnenko, V. (2014) Lisbonizing vs. financializing Europe? The Lisbon Strategy and the (un-)making of the European knowledge-based economy, *Environment and Planning C* 32(1): 108–128.

Birch, K., Levidow, L. and Papaioannou, T. (2010) Sustainable capital? The neoliberalization of nature and knowledge in the European knowledge-based bio-economy, *Sustainability* 2(9): 2898–2918.

Birch, K., Levidow, L. and Papaioannou, T. (2014) Self-fulfilling prophecies of the European knowledge-based bio-economy: The discursive shaping of institutional and policy frameworks in the bio-pharmaceuticals sector, *Journal of the Knowledge Economy* 5(1): 1–18.

Borrás, S. (2003) *The Innovation Policy of the European Union*, Cheltenham: Edward Elgar.

Borup, M., Brown, N., Konrad, K. and van Lente, H. (2006) The sociology of expectations in science and technology, *Technology Analysis and Strategic Management* 18(3/4): 285–298.

Brown, N. (2003) Hope against hype – Accountability in biopasts, presents and futures, *Science Studies* 16(2): 3–21.

Brown, N. and Michael, M. (2002) From authority to authenticity: The changing governance of biotechnology, *Health, Risk and Society* 4(3): 259–272.

Brown, N., and Michael, M. (2003) A sociology of expectations: Retrospecting prospects and prospecting retrospects, *Technology Analysis and Strategic Management* 15(1): 3–18.

Brown, N., Rappert, B. and Webster, A. (2000) Introducing contested futures: From *looking into* the future to *looking at* the future, in N. Brown, B. Rappert and A. Webster (eds) *Contested Futures*, Aldershot: Ashgate, pp. 3–20.

CEC (1993) *Growth, Competitiveness and Employment: The Challenges and Ways Forward into the 21st Century*, Brussels: Commission of the European Communities.

CEC (2002) *Life Sciences and Biotechnology – A Strategy for Europe*, COM(2002) 27 final, Brussels: Commission of the European Communities.

CEC (2005) *Report on European Technology Platforms and Joint Technology Initiatives: Fostering Public–Private R&D Partnerships to Boost Europe's Industrial Competitiveness*, Brussels: Commission of the European Communities.

CEC (2006) *Creating an Innovative Europe*. Report of the Independent Expert Group on R&D and Innovation appointed following the Hampton Court Summit and chaired by Esko Aho, Brussels: Commission of the European Communities. Available at: http://ec.europa.eu/invest-in-research/pdf/download_en/aho_report.pdf

CEC (2007a) *Communication from the Commission on the Mid-Term Review of the Strategy on Life Sciences and Biotechnology*, COM (2007) 175. Brussels: Commission of the European Communities.

CEC (2007b) *Annex I: A Lead Market Initiative for Europe*, SEC (2007) 1729. Brussels: Commission of the European Communities.

CEC (2007c) *Analysis of the effects of a Joint Technology Initiative (JTI) in the area of Innovative Medicines*, SEC (2007) 568, Brussels: Commission of the European Communities.

DG Research (2005) *New Perspectives on the Knowledge-Based Bio-Economy: Conference report*, Brussels: DG-Research.

DG Research (2006) FP7 Theme 2: Food, Agriculture, Fisheries and Biotechnology (FAFB).

DG Research (2007) *Third Status Report on European Technology Platforms at the Launch of FP7*, Brussels: DG-Research.

EFPIA (2004a) *Position paper: Barriers to Innovation in the Development of New Medicines in Europe and Possible Solutions to Address these Barriers*, Brussels: European Federation of Pharmaceutical Industries and Associations.

EFPIA (2004b) *Vision: Innovative Medicines for Europe: Creating biomedical R&D leadership for Europe to benefit patients and society*, Brussels: European Federation of Pharmaceutical Industries and Associations.

EFPIA (2005) *The Innovative Medicines Initiative (IMI): Strategic Research Agenda (draft)*, Brussels: European Federation of Pharmaceutical Industries and Associations.

EFPIA (2009) Innovative Medicines Initiative: EUR 246 million to support public–private research cooperation for a fast development of better medicines, 18 May, press statement.

EFPIA & IMI (2006) *The Innovative Medicines Initiative (IMI): Strategic Research Agenda*, Brussels: European Federation of Pharmaceutical Industries and Associations & Innovative Medicines Initiative.

EU Presidency (2007) *En Route to the Knowledge-Based Bio-Economy*, Cologne: Cologne Summit of the German Presidency.

EuropaBio (2008) Press Release: Biotech industry welcomes European Commission's communication on European Lead Market Initiative, Brussels: EuropaBio.

European Council (2000) *An Agenda of Economic and Social Renewal for Europe* (aka Lisbon Agenda), Brussels: European Council [DOC/00/7].

European Council (2003) *Presidency Conclusions: Brussels European Council* (20–21 March, 2003), Brussels: European Council [8410/03].

Fairclough, N. (2010) *Critical Discourse Analysis: The Critical Study of Language* (2nd edition), London: Pearson.

Felt, U., Wynne, B., Callon, M., Goncalves, M.E., Jasanoff, S., Jepsen, M., Joly, P.-B., Konopasek, Z., May, S., Neubauer, C., Rip, A., Siune, K., Stirling, A. and Tallacchini, M. (2007) *Science and Governance: Taking European Knowledge Society Seriously*, Brussels: Commission of the European Communities.

Genewatch UK (2010) *Bioscience for Life? Who decides what research is done in health and agriculture?* Available at: http://www.genewatch.org/

Georghiou, L. (2001) Evolving frameworks for European collaboration in research and technology, *Research Policy* 30(6): 891–903.

Gibbon, P. and Ponte, S. (2008) Global value chains: From governance to governmentality, *Economy and Society* 37(3): 365–392.

Gillespie, I. *et al.* (2007) *The Innovative Medicines Initiative: Assessment of Economical and Societal Effects*, Brussels: Commission of the European Communities. Available at: http://imi.europa.eu/docs/imi-ia-report-032007_en.pdf

Gottweis, H. (1998) *Governing Molecules*, Cambridge MA: MIT Press.

Guice, J. (1999) Designing the future: The culture of new trends in science and technology, *Research Policy* 28(1): 81–98.

Hedgecoe, A. (2003) Terminology and the construction of scientific disciplines: The case of pharmacogenomics, *Science, Technology and Human Values* 28(4): 513–537.

Hedgecoe, A. and Martin, P. (2003) The drugs don't work: Expectations and the shaping of pharmacogenetics, *Social Studies of Science* 33(3): 327–364.

Helen, I. (2004) Health in prospect: High-tech medicine, life enhancement and the economy of hope, *Science Studies* 17(1): 3–19.

IMI (2006) The Innovative Medicines Initiative: Keys for Success – Industry Input. Available at: http://www.imi-europe.org/Lists/IMIPublicationDocuments/20070309_IMI_Keys_for_Success%20Final.pdf

IMI (2007) Intellectual Property Policy, Brussels: Innovative Medicines Initiative. Available at: http://imi.europa.eu/docs/imi-ipr-policy01august2007_en.pdf

IMI (2008) *The Innovative Medicines Initiative (IMI) Research Agenda: Creating Biomedical R&D Leadership for Europe to Benefit Patients and Society*, Brussels: Innovative Medicines Initiative.

Jasanoff, S. (ed.) (2004) *States of Knowledge: The Co-production of Science and Social Order*, London: Routledge.

Jasanoff, S. and Kim, S.-H. (2009) Containing the atom: Sociotechnical imaginaries and nuclear power in the United States and Korea, *Minerva* 47: 119–146.

Jessop, B. (2005) Cultural political economy, the knowledge-based economy, and the state, in A. Barry and D. Slater (eds) *The Technological Economy*, London: Routledge, pp. 144–166.

Jessop, B. (2009) Cultural political economy and critical policy studies, *Critical Policy Studies* 3: 336–356.

Levidow, L., Birch, K. and Papaioannou, T. (2012) EU agri-innovation policy: Two contending visions of the knowledge-based bio-economy, *Critical Policy Studies* 16(1): 40–66.

Loconto, A. (2010) Sustainability performed: Reconciling global value chain governance and performativity, *Journal of Rural Social Science* 25(3): 193–225.

Merton, Robert K. (1968) *Social Theory and Social Structure*, New York: Free Press.

Nightingale, P. and Martin, P. (2004) The myth of the biotech revolution, *Trends in Biotechnology* 22(11): 564–569.

OECD (2005) *The Bioeconomy to 2030: Designing a Policy Agenda*, Paris: Organisation for Economic Co-operation and Development.

OLS (2009) *Life Sciences Blueprint: A statement from the Office for Life Sciences*, London: Office for Life Sciences, Department of Business, Innovation and Skills.

Peck, J. (2011) Geographies of policy: From transfer-diffusion to mobility-mutation, *Progress in Human Geography* 35(6): 773–797.

Ponte, S. and Birch, K. (2014) Introduction: Imaginaries and governance of biofueled futures, *Environment and Planning A* 46(2): 271–279.

Ragan, C.I. (2007) Pre-competitive R&D: Applying science along the whole value chain from early discovery to pharmacovigilance, EFPIA presentation at Future Pharma UK 2007. Available at: http://www.imi-europe.org/Lists/IMIEventAttachments/Future%20Pharma%202007.pdf

Rappert, B. (1999) Rationalising the future? Foresight in science and technology policy co-ordination, *Futures* 31(6): 527–546.

Sanz-Menéndez, L. and Borrás, S. (2000) *Explaining Changes and Continuity in EU Technology Policy: The Politics of Ideas*, Madrid: Unidad de Políticas Comparadas (CSIC) Working Paper 00-01. Available at: http://digital.csic.es/bitstream/10261/1490/1/dt-0001.pdf

Smith, K. (2005) Changing economic landscape: Liberalisation and knowledge infrastructures, *Science and Public Policy* 32(5): 339–47.

Surel, Y. (2000) The role of cognitive and normative frames in policy-making, *Journal of European Public Policy* 7(4): 495–512.

Valiverronen, E. (2004) Stories of the 'Medicine Cow': Representations of future promises in media discourse, *Public Understanding of Science* 13: 363–377.

van Lente, H. and Rip, A. (1998) The rise of membrane technology: From rhetorics to social reality, *Social Studies of Science* 28(2): 221–254.

6 Innovation financing in the global life sciences

Introduction

As the last chapter indicated, any examination of the life sciences requires an analysis of the discourses or imaginaries, especially as this relates to future visions, expectations and promises. As I outlined, these imaginaries play an important performative role on a number of levels. For one, they provide the means to enrol a range of stakeholders in an often uncertain undertaking. It is not, for example, clear whether life sciences research will actually lead to a commercial product or service; and in the vast majority of instances this is the case, research ends up being just that (Scannell 2015; Mittra 2016). For another, imaginaries provide direction for policy-makers in planning future policies and changing policy frameworks and institutions (Birch *et al.* 2010, 2014). As important, especially in the context of this chapter, imaginaries play a critical role in the financing of the life sciences, and innovation more generally. A number of scholars in STS, for example, have conceptualized this role in terms of 'promissory value' or 'promissory economies' (e.g. Sunder Rajan 2006, 2012; Waldby and Mitchell 2006; Cooper 2008; cf. Birch 2012; Birch and Tyfield 2013). It is not my intention to engage with this largely theoretical literature here however; rather I focus on the more grounded work on the (global) financing of the life sciences and the implications of finance for the life sciences.

The financing of the life sciences is an important topic when it comes to the development of the sector and its implications for regional development. Early research in this area (e.g. Powell *et al.* 2002) stressed the co-location of venture capital (VC) and new 'biotech' firms in regional economies like California. Others like Cooke (2007) highlight the importance of such intermediary social actors in regional innovation systems. However, despite this research, there is an emerging literature that is critical of the assumption that finance – especially VC – plays an unproblematic role in life sciences innovation. An example of this is the work of Gary Pisano (2006), especially his book *Science Business*. Since then a number of other scholars have highlighted the implications of the growing *financialization* of the life sciences (e.g. Andersson *et al.* 2010; Lazonick and Tulum 2011; Hopkins *et al.* 2013; Styhre 2014, 2015; Birch forthcoming). Finance and financialization have also become major topics in debates about uneven development and regional economies, especially since the global financial crisis (e.g. Pike and Pollard 2010; Coe *et al.* 2014; Sokol 2015).

My aim in this chapter is to outline and analyze the financing of the global life sciences in order to highlight a range of problematic issues that such financing, and its ups and downs (e.g. volatility), has for life sciences innovation. In order to do so, I start by discussing the concept of financialization, which represents the notion that finance has come to dominate all other aspects of economic activity (e.g. production). I then turn to the global life sciences drawing on secondary data to illustrate how the global life sciences sector is financed and what this means for innovation. This secondary data analysis sets up the rest of the empirical analysis, which draws on a series of in-depth interviews with life sciences financiers and investors. Although these informants are drawn from a UK context, they are involved in financing life sciences firms around the world. It is important to remember that finance, in any context, goes beyond regional and national scales, integrating a range of social actors in the innovation process as a result.

Finance, financialization and the life sciences

Finance has become an increasingly important topic as the success of innovation has been specifically aligned with the financing of research and innovation in policy circles. As Chapter 5 showed, policy discourses are influential in shaping policy-making and reshaping policy frameworks. In the context of policy debates about economic growth and development, policy-makers in Europe, for example, have focused on the characteristics of supposedly *more* innovative countries like the USA (e.g. European Council 2000; CEC 2010). One thing policy-makers identified as distinct about the USA was its financial market, which was seen as a central component of the USA's innovation ecosystem (Rosamond 2002). Consequently, European policy discourse has somewhat fetishized the role of finance in innovation. For example, the 2000 *Lisbon Agenda* stated that:

> Efficient and transparent financial markets foster growth and employment by better allocation of capital and reducing its cost. They therefore play an essential role in fuelling new ideas, supporting entrepreneurial culture and promoting access to and use of new technologies. It is essential to exploit the potential of the euro to push forward the integration of EU financial markets. Furthermore, efficient risk capital markets play a major role in innovative high-growth SMEs and the creation of new and sustainable jobs.
>
> (European Council 2000)

This discourse has driven attempts by policy-makers to transform European financial markets and regulations, especially as a way to stimulate and support innovation (Grahl 2011; Cleeton 2012; Deeg 2012). However, work by myself and my colleague Vlad Mykhnenko has questioned whether this financialization of the European economy has actually led to the emergence of a 'knowledge-based economy' in Europe, rather than simply contributing to the problematic rise of finance (Birch and Mykhnenko 2014). Our conclusion was that it had not; instead of promoting high-tech sectors, for example, the Lisbon Agenda period actually entailed a far more significant expansion of the finance sector. In this regards, finance cannot be simplistically equated with innovation; nor can venture capital or capital markets be identified as key drivers of innovation.

It is not surprising that finance has become a major research topic in recent years, following the 2007–2008 global financial crisis (GFC). Similarly, it is not surprising that the impact of finance on the wider economy and society – or what has been termed *financialization*, which I discuss below – has become an increasingly debated issue. Obviously, a number of academics had been writing about finance before the GFC, some well before; examples include people like Boyer (2000), Lazonick and O'Sullivan (2000) and Krippner (2005). In this early and later literature, 'finance' is generally used to refer to the financing of business enterprises, which includes equity investment (i.e. shares), debt (i.e. loans), and other financial instruments (i.e. securities). According to scholars like Lazonick and O'Sullivan (2000), finance and financing considerations have increasingly shaped management and strategic decisions since the 1970s, especially in terms of the pursuit of shareholder returns at the expense of other activities (e.g. R&D investment).

This growing influence of finance led scholars to conceptualize this change as a process of 'financialization'. In the literature, financialization is defined in a number of similar ways: Krippner (2005: 174) uses it to refer to 'a pattern of accumulation in which profits accrue primarily through financial channels'; Leyshon and Thrift (2007: 102) refer to the 'growing power of money and finance within economic life'; similarly, Coe *et al.* (2014: 763) state that it is about 'the growing power of financial markets and institutions in national economies'; and Pike and Pollard (2010: 30) argue that it reflects 'the growing influence of capital markets, their intermediaries, and processes'. As these definitions imply, financialization is generally used to define processes that operate in particular sectors (e.g. banking) and particular geographies (e.g. national scale). That being said, it is also conceptualized in ways that go beyond these dimensions; as a result, it can be used to analyze the global financing of the life sciences.

Finance plays an important role in life sciences innovation. Over the last couple of decades, a number of scholars have identified specific types of relationship between finance and the life sciences. As mentioned in the introduction, research on the regional dimensions of finance and financing the life sciences has tended to focus on the co-location of finance (e.g. venture capitalists) and research and innovation (e.g. 'biotech' firms); examples here include Powell *et al.* (2002) and Cooke (2007). In my own work on *knowledge-based commodity chains* (e.g. Birch 2008, 2011; Birch and Cumbers 2010), which I discussed in Chapter 4, this particularity of financing the life sciences can be attributed to a number of characteristics of life sciences research and innovation like the high asset specificity of research, uncertainty in the innovation process, and the length of time needed in product development and trials. Other work in this area stresses a number of related aspects to this relationship, including some specific implications of the financialization to the life sciences.

First, Pisano (2006) argues that the life sciences are characterized by a monetization of intellectual property (IP) through licensing, royalties, etc., which provides the basis for the capitalization of life sciences firms – that is, their public market valuation. As a consequence of this strategy, Pisano argues, life sciences firms have little incentive to actually develop new products and services as their IP activities generate income. Second, Andersson *et al.* (2010) argue that the life sciences are characterized by a 'financialized business model' in which life sciences firms are more focused on increasing their market capitalization than developing new products. This strategy results from a 'relay-like' financing process in which ownership (i.e. shares) are passed from one investor to the next, necessitating constantly increasing share value (see Hopkins 2012). Third, Mirowski (2012) has explicitly referred to life sciences firms as 'ponzi schemes' since a vast majority of them never develop a product, meaning that investors are simply aiming to cash in before the firm goes bust.[1] Finally, in my own work I argue that the life sciences sector is characterized by a process of *assetization* in which investors (e.g. VC) primarily seek to turn IP and firms into capitalized property (i.e. assets) that can be sold, rather than support product development (Birch 2015, forthcoming).

As this research should illustrate, the relationship between finance and the life sciences is fraught with problems. As these issues relate to regional development, the role of finance in innovation illustrates the extent to which regions cannot be considered as financially 'bounded' territories; rather, as Coe *et al.* (2014) point out, regional (and national) borders are financially porous. As illustrated in Chapter 3, for example, UK life sciences firms did not draw predominantly on local finance; in fact, my research shows that both national and international finance were generally more important to their innovative activities (see Figure 3.4). This finding applies to 'clustered' firms as much as 'non-clustered'. Similarly, some of my other research has shown that different kinds of finance have different geographies (e.g. Birch 2011). On the one hand, institutional investment (e.g. pension, insurance and mutual funds) is liable simply to extract value from regions; while, on the other hand, business angel investment ends up being recycled through localized economies, meaning that value does not end up going elsewhere. Others, like Sokol (2015), have highlighted the highly uneven geographical dimensions of financialization suggested by this research. For the chapter, however, my intention is to focus on how finance impacts on innovation at the firm-level, especially in terms of driving business and research strategies, and come back to the more specific issues relating to regional development in the following chapter.

The global financing of the life sciences

Methodological note

In previous research projects between 2002 and 2008 – which I discuss in Chapters 3 and 4 – I had asked general questions about the financing of life sciences firms, both in surveys and interviews. However, these projects did not explicitly focus on finance or the process of financialization, nor their implications for life sciences innovation. As a result, I undertook new research in 2011 and 2012 to examine the financialization of the life sciences, especially as it related to the implications of the global financial crisis (GFC) on the UK life sciences. I have since then extended my research in this area to Canada, but have not completed this research yet. The research reported here involves the analysis of secondary data and of in-depth interviews with 13 British-based investors, financiers and brokers who focus on the life sciences sector. These interviews cover a range of questions dealing with the impacts and implications of the GFC on life sciences firms, especially on their business and research strategies. In this section, I start by presenting the secondary material on the global financing of the life sciences, before then focusing on the primary empirical material from the interviews.

The global life sciences

Until very recently, the global 'biotech' industry – as defined by a longstanding series of Ernst & Young reports – was not profitable (see Figure 6.1). In fact, up until the late 2000s the revenues of the global biotech industry had *never* been positive; to put that in perspective, from the founding of Genetech in 1976 until 2009 – over 30 years – the whole global sector had not returned an aggregate profit (EY 2015). That does not mean that individual life sciences firms have not been profitable; some firms like Amgen and Genentech have been profitable and represent a majority of overall revenues, profits and share valuations. Until relatively recently, then, the global industry was being driven by the *promise* of future earnings, reflected in share valuations but not, generally speaking, in the development of new products. As a result, returns on investments have been variable, to say the least. For example, Pisano (2006) argues that investor returns on publicly-listed life sciences firms have been generally poor; moreover, Hopkins *et al.* (2013) argue that some venture capitalists have made significant returns on their investments, but a significant number have not. Consequently, it is important to understand that much of the financial value in the life sciences is on paper only (i.e. share valuations).

Only a few large, publicly-listed firms represent the lion's share of global revenues, profits, and rising share values. Looking at a range of secondary data on global revenues and profitability of publicly-listed firms illustrates the extent to which the global industry is dominated by only a few firms – see Figure 6.2. As these data show, the revenues of large life sciences firms rose from near US$30 billion in 2004 to near US$65 billion in 2011, while the revenues of all the other sized firms had slight or no real rise in the same period. Similarly, during the same period the profits of large firms rose from near US$5 billion to near US$14 billion, while all other sized firms ended up losing money in 2011. Notably, these data on large firms represent the revenues and profitability of only between 10 and 15 large firms at any point in the time period. By 2011, for example, there were 15 large firms constituting *two-thirds* of global revenues and *all* global profits. It goes without saying that these few firms – out of a total around 4,500 worldwide (EY 2015) – are the firms with successful blockbuster drugs on the

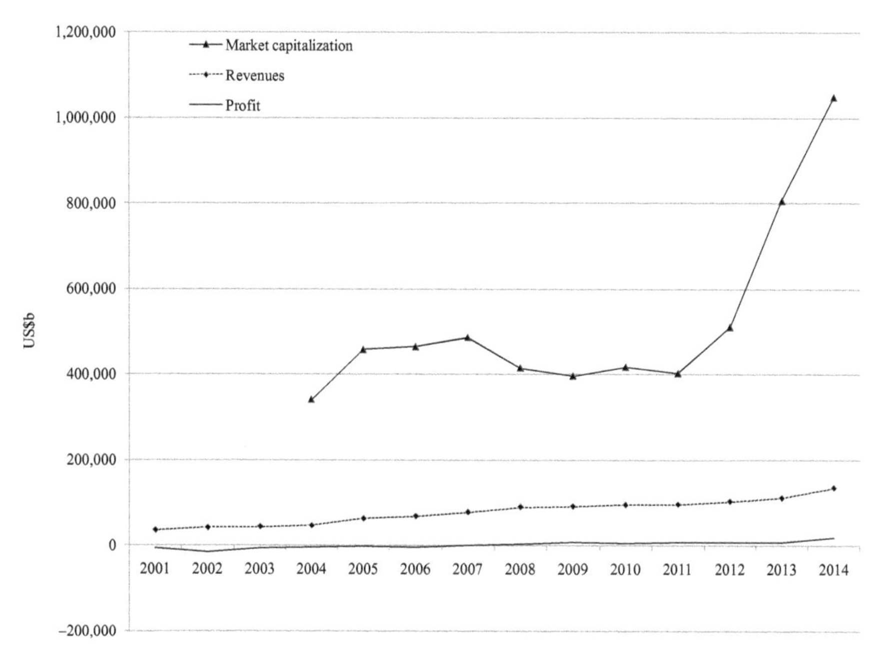

Figure 6.1 Global biotech industry: market capitalization, revenues and profit (US$b)

Source: Lähteenmäki and Lawrence (2005); Lawrence and Lähteenmäki (2008, 2014); Huggett *et al.* (2009, 2010, 2011); Huggett and Lähteenmäki (2012); Huggett (2013); Morrison and Lähteenmäki (2015); reproduced with the permission of this book's author.

market. The reason I emphasize this point here is to highlight how lopsided the life sciences sector is, a topic I will come back to in the next chapter.

If I wanted to make a simplistic analysis of these trends, I might suggest that investment in the life sciences sector is highly uncertain, yet potentially highly profitable. For example, the fact that 15 firms make two-thirds of the global revenues and all the profits implies that financing is driven by such a high-risk, high-reward strategy. However, this suggestion obscures a more interesting process. There is considerable value – financial and political-economic – across the life sciences sector, in the unprofitable and smaller firms as well as the few large ones. It is important to understand this value. As some of the scholars I mention above point out (e.g. Andersson *et al.* 2010; Hopkins 2012; Hopkins *et al.* 2013; Birch forthcoming), financing the life sciences involves more than a single investment in a firm. Rather, financing involves a series of stages, each of which entails a set of decisions about future potential (e.g. milestones) and then the management of research and business strategies around those decisions – see Chapter 7 for some of the implications of this. As such, it is as important to analyze what value is located in firms themselves (i.e. their assets and equity) and in any mergers and acquisitions (e.g. trade sales), as it is to analyze any returns on investment. In some ways, financing the life sciences has ended up less dependent on the ultimate

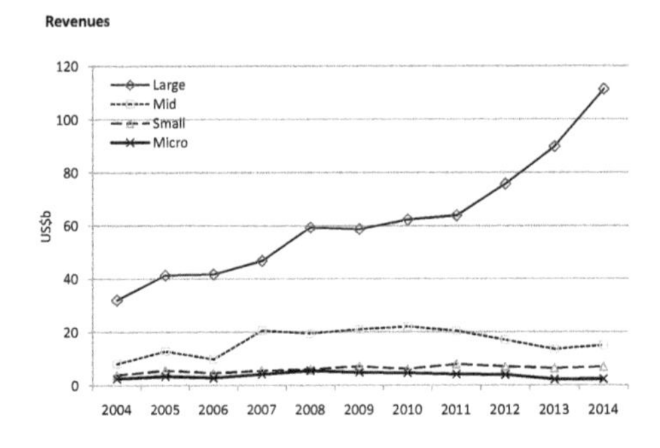

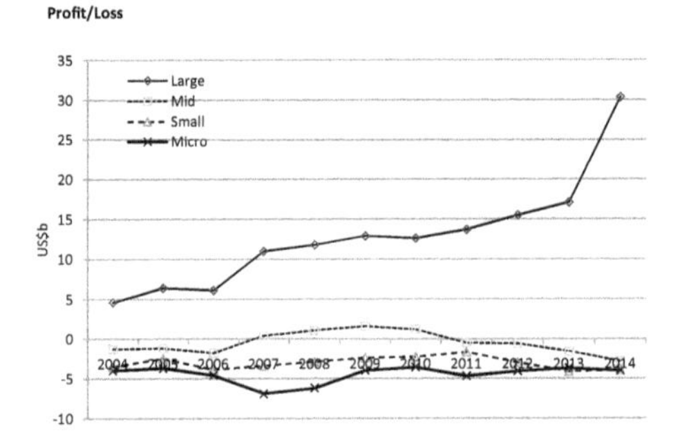

Figure 6.2 Global 'public' biotech industry by size of firms: revenues (top graph) and profit/loss (bottom graph) (US$b)

Source: Lähteenmäki and Lawrence (2005); Lawrence and Lähteenmäki (2008, 2014); Huggett *et al.* (2009, 2010, 2011); Huggett and Lähteenmäki (2012); Huggett (2013); Morrison and Lähteenmäki (2015); reproduced with the permission of this book's author.

development of a marketable product than on a 'relay' between investors as they pass ownership rights on to one another, until some are either left with (most often) nothing or (sometimes) everything (Andersson *et al.* 2010; Birch forthcoming).

Financing the life sciences

As the above context should demonstrate, it is important here to understand how the life sciences are financed before considering the implications of financialization for life sciences innovation. Life sciences firms are dependent on what some informants called 'patient capital' considering the length of time it takes to develop products, which can range between 10 and 20 years (Birch and Cumbers 2010; Hopkins *et al.* 2013). However, most financial investments, including supposedly risk-taking venture capital, does not work on such long-term time horizons; instead, VC investments are largely driven by the need for relatively short-term returns to pay back investors. As one informant explained, 'I think that anybody who's got a limited life fund [e.g. VC] ultimately wants a return ... you know a capital based exit' (VC Investor K). It is unsurprising then, and as mentioned in earlier parts of this book (e.g. Chapter 4), that the public sector plays a critical role here in providing pre-seed and seed investment (e.g. Birch 2011). In her recent book, *The Entrepreneurial State*, Mariana Mazzucato (2013) argues that government funding is vital for kick-starting innovation since private investors are risk averse, especially in highly uncertain sectors like the life sciences. The traditional *imaginary* – see last chapter – of the dynamic, venture-backed life sciences firm is basically a myth in this sense. As Hopkins (2012) and others demonstrate, most life sciences firms are actually more reliant on other forms of investment. The form of investment varies quite significantly depending on the stage in the innovation process. Earlier stages are defined by the so-called '3Fs' – or 'family, friends and fools' as one informant told me in 2008. Alongside the 3Fs, government funding provides an important source of seed and early-stage financing (Birch 2011).

Life sciences financing is complex. Even the notion of a financing relay, which I discussed above, belies the complicated financing arrangements in the life sciences. There are: (1) different types of financing; (2) different financing actors; and (3) different implications for each of these financing arrangements. In a recent handbook for life sciences managers, Hopkins (2012: 137) provides a useful diagram of this variety. Pre-seed and seed financing is usually provided through the public funding of basic research and early commercialization, whether from government directly or through public institutions like universities. Many regional governments have financing schemes for pre-seed and seed level funding; for example, Scottish Enterprise established the *Scottish Seed Fund* – alongside the *Co-investment Fund* and *Venture Fund* – to finance firms in high-tech sectors (Birch 2011). As firms move into early-stage funding, there is an increasing reliance on business angels, venture capital and, especially since the GFC, corporate venturing (Hopkins 2012). The latter include venture funds established by large pharmaceutical firms, such as SR One of GlaxoSmithKline. At this early stage, firms are unlikely to have developed any marketable products yet; however, they may be able to demonstrate the disruptive potential of their technologies. As one VC informant explained:

> ... our focus is to find companies which are developing a disruptive technology where the disruption is sufficiently apparent that it will trigger a strategic acquisition by an incumbent in the market.
>
> (VCT Investor F)

Later stage financing is largely driven by the search for exit by existing investors, whether this is through a trade sale (as suggested in the quote above) or initial public offering (IPO). Other forms of financing at this stage include partnerships (e.g. out-licensing, royalty arrangements) and debt, which becomes more accessible as firms get closer to launch of a product. As Hopkins (2012) illustrates, though, this chronology is, by no means, linear or straightforward; financing methods often overlap, sometimes contradict, and create tension with existing innovation strategies.

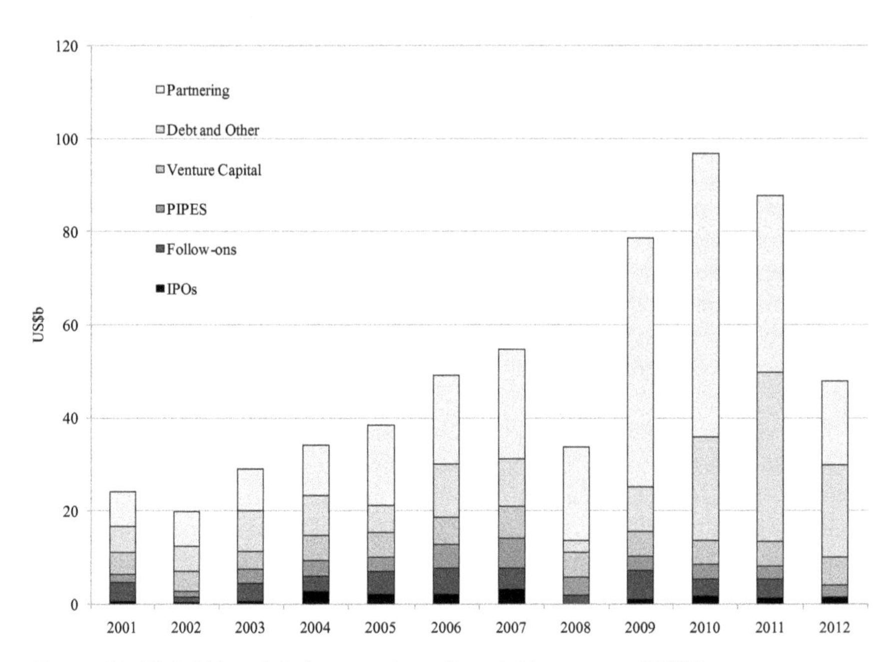

Figure 6.3 Global biotech industry: private financial investment (US$b)

Source and Notes: Lähteenmäki and Lawrence (2005); Huggett and Lähteenmäki (2012); Yang (2014); data for partnerships for the years 2009–2012 includes global deals. Reproduced with the permission of this book's author.

My aim in this discussion has been to unpack the view that life sciences innovation is constituted by risk-taking VC (e.g. OLS 2011); it clearly is not. As the data in Figure 6.3 shows, VC financing represents only a small proportion of total life sciences financing. Other forms of financing are far more significant. This data covers global life sciences financing from 2001 to 2012, and is derived from regular reports in the scientific journal *Nature Biotechnology*. Although there are some issues with data, it is a useful illustration of the diversity of global financing. It shows how important partnering income and debt are to financing life sciences innovation, and how relatively unimportant VC and IPOs are. It does not show, however, the shift towards trade sales (i.e. mergers and acquisitions) in the life sciences, especially since the GFC. Current research on the financing of the life sciences shows that these sorts of 'exit' events have come to dominate innovation and business strategies (Hopkins 2012; Styhre 2014; Birch forthcoming), almost to the exclusion of IPOs. In part this is the result of how financing is organized in that it has to return capital to the investor at the end of a certain time period, and cannot simply maintain the investment indefinitely. And here, trade sales are considered to be 'clearer, cleaner, quicker' (VCT Investor F) than other exits.

Finance in the life sciences comes in many forms (Hopkins 2012). Evidently, different forms of finance involve different valuation knowledges, practices and processes (Birch forthcoming). That being said, one obvious trend around the

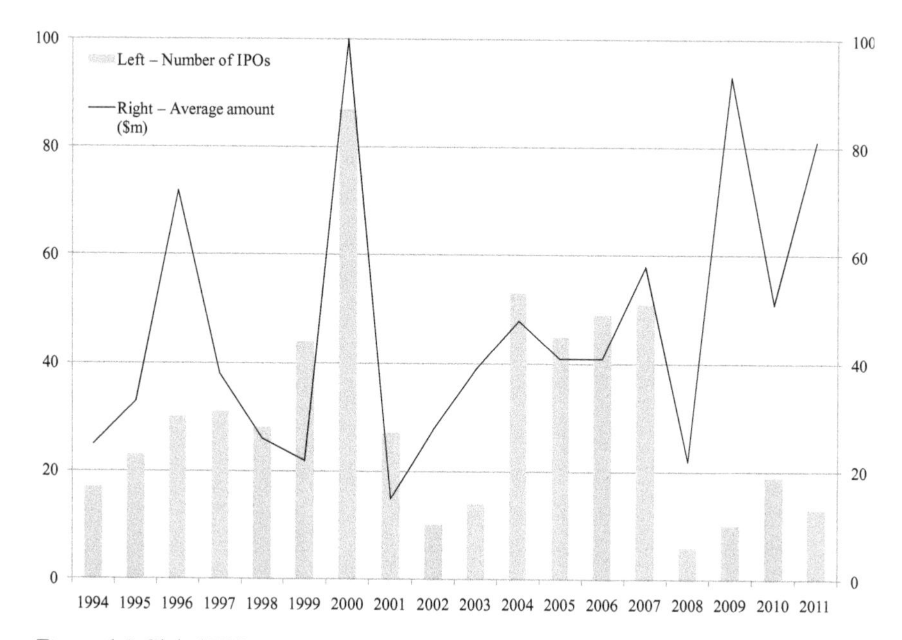

Figure 6.4 Global IPOs

Source: Lähteenmäki and Lawrence (2005); Huggett and Lähteenmäki (2012); reproduced with the permission of this books' author.

world after the GFC started in 2007–2008 was the collapse of the public capital markets as an exit for life sciences investors. Not only did life sciences private financing generally nose-dive around the world, as Figure 6.3 should illustrate, the number of IPOs worldwide fell off a cliff in 2008. This consequence of the GFC had not reversed itself by 2011, when I first started looking at the financing of the UK life sciences (see Figure 6.4). The effect of this collapse of IPOs as an exit opportunity on life sciences innovation is ambiguous, especially when factoring in the implications for regional development. As argued in the literature (e.g. Hopkins 2012; Birch forthcoming), IPOs did not represent the main way that value was realized in the life sciences sector, before the GFC and definitely afterwards. And this can be a benefit to regional development since value is not sucked out of local economies to distant financial markets, usually governed and managed in global cities like London, New York and Toronto (Birch 2011). Instead, trade sales can often help recycle localized investments (e.g. business angels, regional VC) back into regional economies as investors receive their capital returns and look for new investment opportunities. I will come back to this in the next chapter.

The impact of the 2007–2008 global financial crisis on the UK life sciences

The global financial crisis (GFC) has been framed as a deeply problematic event for the life sciences sector. For example, an August 2011 story in *The Guardian*

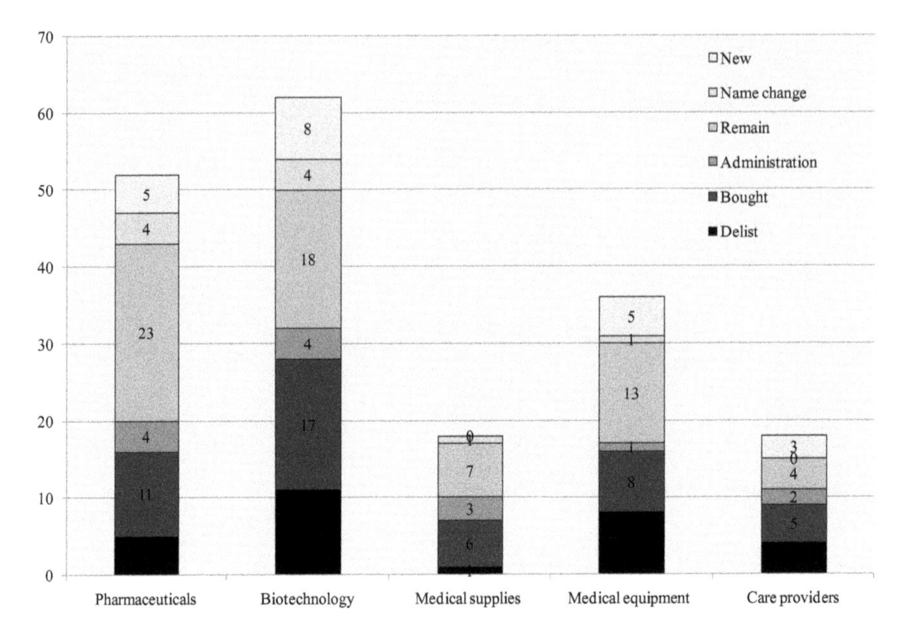

Figure 6.5 Change in UK life sciences firms on public markets (2007 vs. 2011)
Source: London Stock Exchange.

newspaper, titled 'Britain's biotech stars fade away', claimed that 'More than a third – 29 companies – of the [UK] listed biotech sector have gone bust since 2008'. On top of this, it went on, '10 are on the brink'. The article implied that the UK life sciences were facing a serious setback as the result of the GFC; new financing had dried up, as had public markets as an exit, alongside which investors were withdrawing their existing investments. This situation was not limited to the UK, moreover, as the data outlined above shows.

Despite this report in *The Guardian*, however, it was not always clear whether the GFC had such a devastating effect on the UK life sciences. In particular, secondary data on life sciences firms listed on UK capital markets showed something different happening, something more interesting. These capital markets include both the Alternative Investment Market (AIM) and mainstream London Stock Exchange (LSE). I collected data on the 'life sciences', which covers a broad range of firms from 'pharmaceutical', 'biotechnology', 'medical devices', 'medical supplies', and 'healthcare' sectors. The analysis here is not, therefore, limited to 'biotechnology', although it includes that sector. I collected secondary data on publicly-listed firms on two dates – 1 January 2007 and 31 December 2011 – in order to explore what happened as the result of the GFC. This information is contained in Figure 6.5.

As these data illustrate, the life sciences sector in the UK experienced a major shake-out as a result of the GFC, although it did not necessarily happen in the way presented in the 2011 *Guardian* newspaper story mentioned above. The data in Figure 6.5 show that between January 2007 and December 2011, nearly 32 'biotechnology' firms left the UK public capital markets and 8 joined them. Of the businesses that left, however, most were bought ($n = 17$) or delisted ($n = 11$); only a small number ($n = 4$) actually went bankrupt. Overall, covering the whole 'life sciences' sector, there was a decline of around 42 per cent in number of publicly listed life sciences businesses between 2007 and 2011. A significant number, but again most of these businesses were bought or delisted. The primary reason that delisting was so common during and immediately after the GFC was because small, high-tech firms find it very difficult to generate 'trading data' (e.g. new patent applications, new clinical trial results, etc.) on a regular basis, meaning that their share values tend to decline gradually since they are not being traded (Birch forthcoming). Consequently, the costs of staying on the public market came to outweigh the benefits.

Changing strategies in the life sciences

In light of the secondary data above, it is critical to understand what impacts the GFC has had on life sciences innovation, especially in terms of the business strategies pursued by firms. In order to examine these impacts, I analyze a number of in-depth interviews with informants drawn from the financial world, specifically experts in financing the life sciences. I discuss, in turn, a number of key issues or themes that arose from these interviews relating to the impacts of the GFC on: innovation promises; capital availability; the financial ecosystem; exit opportunities; and R&D strategies.

Innovation promises

One of the reasons I included a chapter on imaginaries in this book – see Chapter 5 – is to illustrate how important future visions, narratives and promises are to the life sciences (if not all sectors). Their notion that they have a performative effect – that is, directly shaping the outcomes imagined in those promises – is contestable, however, as I and others have argued elsewhere (e.g. Birch 2007; Ponte 2009). Rather, these promises can be considered as an example of 'virtualism' (Miller 2002); by this I mean that promises, visions, narratives, etc. (i.e. discourse) does have a effect, but it is not necessarily a material effect or an effect as imagined. In relation to the life sciences, a number of informants, for example, stressed that it has been over-sold or hyped; for example, one argued that:

> I don't think there's going to be very much change [as the result of the GFC]. I think that the market is actually reasonably efficient at the moment, you know. I think there was a sort of great deal of interest in life sciences ten years ago because there was this belief that universities contained all these potential therapies which were not being exploited and they should be exploited

> and, you know, that they would be the start of some great revolution in the life sciences worldwide. And I think that's actually a chimera.
>
> (VC Investor K)

Another informant reinforced this comment:

> ... if you take much of what was being done five or seven years ago with life science businesses, it was research. It wasn't business, it was research.
>
> (VC Specialist Investor C)

Others informants, while offering similar diagnosis of this innovation promise, even went so far as to attempt some form of quantitative analysis of the process:

> I did some work in about 2002 or '03, I would think, in a different place. And I was interested to discover that there was absolutely no correlation between what stage the molecules were at and how many molecules there were on the share price. It was all to do with the perception of the prospect. And really, that ... a lot of these businesses went bust.
>
> (BA Syndicate Investor L)

As these quotes suggest and others have theorized (e.g. Pisano 2006; Andersson et al. 2010; Mirowski 2012; Birch forthcoming), the life sciences sector is configured by finance, probably to a greater extent than research. It is, in this sense, thoroughly financialized. The implications of this are important to consider, since they impact on both business and research strategies. For example, one informant commented that: 'So I think the companies that usually invest in a reasonably small [firm] ... they're never perfect, that's why we sell them, they're never perfect' (VC Investor B). It does not matter, in this context, whether innovation promises are realized or not, since the main investor strategy is to sell a firm before it (fails to) develop a product (Pisano 2006).

Capital availability

All the secondary data above, whether it is global or UK data, suggests that financing for the life sciences has declined significantly as the result of the GFC. This is confirmed by a range of informants who noted that it is a lot more difficult to get funding after the GFC, although some questioned whether this was any different for any other sector than the life sciences (e.g. VCT Investor E). Several informants made comments to this effect, for example:

> *VC Investor B*: I think there are quite a few VC funders struggling in the last few years with our portfolios and struggled for VC funds.
>
> *VC Specialist Investor C*: Well, it's tougher [to raise money]. I think most of the money's gone out of the market.

> *Venture Debt Investor D*: Obviously what's happened to [Firm X] is a good example of how funding to the funders, the sector, has been switched off. Other people, and weren't the only casualty.

Following the GFC then, money was 'turned off', although this happened for a variety of reasons that informants did not directly blame it on the GFC. For example, one noted that the GFC happened at a time when many VC funds were coming to the end of their funding cycle (VCT Investor E). This meant that a number of VC funds had to return capital to their investors (e.g. pension funds, insurance funds, etc.), which meant finding an immediate exit opportunity, and then could not find new investment to create new VC funds. This has had an important knock-on effect that I discuss next. Before that, though, it is worth noting that different financing mechanisms, as outlined by Hopkins (2012) and others, have been impacted in different ways by the GFC; for example, public funding from government has not necessarily declined, while VC funding has. However, the capital markets play another important role in financing of the life sciences, one which is often ignored but relates to valuation practices (Birch forthcoming).

Financial ecosystem

While money is important, it is only part of the financing puzzle. Some of my earlier research on the Scottish life sciences, discussed in Chapter 4, had highlighted the importance of financial actors and the wider financial ecosystem to life sciences innovation (Birch 2011; also see Powell *et al.* 2002; Cooke 2007). As a result of the GFC, one informant noted that there were 'probably *fewer fund managers* with less money today that there were prior to 2008' (VC Investor K; emphasis added). Another claimed that there were far fewer specialist financiers (e.g. specialist VC) in the life sciences as a result of the GFC: 'So when I first joined in, there were probably ten established players in the market. So they went down to about four from about six or seven. So it waxed and waned' (VC Specialist Investor C). The informant explained that a number of new players came into the market in 2005–2007, including hedge funds, public investors, and others, but that this changed since the start of the GFC. As a result, there are far fewer life sciences specialists in the capital markets after the GFC. While the importance of capital availability might seem obvious, the transformation of the life sciences financial ecosystem, with the loss of specialist financiers, analysts, brokers, etc., helps to explain the finding that firms were more likely to delist than go bankrupt as a result of the GFC (see above). The reason is fairly simple: financing necessitates more than capital, it also involves a range of knowledges, practices and actors so that *valuations* can actually be made (Birch forthcoming). Without specialist financiers, the amount of capital investment is likely to be lower as generalists are more risk-averse; without specialist analysts, the valuation of investments are likely to be lower as generalists do not have the knowledge to make value judgements; and without specialist brokers, life sciences firms cannot get listed on capital markets at levels required to make it worthwhile. This ecosystem is particularly critical because capital markets are generally constituted by large, generalist institutional funds (e.g. pensions, insurance, mutual) with little knowledge of the life sciences sector. As one informant noted, these institutional investors 'are the market' to all intents and purposes (Broker G). Any change to the ecosystem, then, has an impact on how investors seek to realize returns on their investments.

Exit opportunities

An important difference between before and after the GFC is the change in investment strategies by financiers (e.g. VC) and firms themselves. It is likely that this has had significant impacts on research strategies as well, which I turn to below. Before that, however, it is important to consider whether the GFC altered the financing 'relay' described by Andersson *et al.* (2010) and others. In the language of financiers, an 'exit' is how they realize the value of their investment and return capital – minus their management fees – to investors (usually institutional investors). For example:

> Well, about 10 years ago there was an awful lot of small technology businesses in the UK, on AIM [Alternative Investment Market] ... 10 years ago, if you were investing in submissions, you would have routes to exit with one trade sale to [one] strategic to [one] float on AIM, so for financial buyer.
>
> (VC Investor B)

As mentioned already, certain exits are fetishized in the discussion of the life sciences, specifically IPOs (e.g. OLS 2011). However, and as mentioned already, IPOs dried up after the GFC (see Figure 6.4), meaning that they no longer represented a viable exit opportunity; most of the informants highlighted this closing of public markets as an exit (e.g. VCT Investor E). After 2007–2008, then, the only real exit options left in the life sciences sector were trade sales (most common) or strategic investing; I discuss the latter below, so focus on the former here. Three informants summed this up:

> *VC Investor B*: So the AIM float exit is gone really, which leaves sales and strategic all to financial buyers.
>
> *Venture Debt Investor D*: Certainly IPO markets, if that hasn't closed significantly down on where it should be or where it was ... I think where there is certainly a lot more activity, without question, is trade sales and trade buyers.
>
> *BA Syndicate Investor L*: And the listing market was incredibly volatile. It would change from month to month, the appetite of institutions to invest, which meant the institutions were getting edgy about what was going to happen, I think. So, really, we kind of abandoned it [AIM].

There are at least two reasons for this change, the first being the disruption of the financing ecosystem I mentioned above; the other reason follows on from this. IPOs and trade sales involve different valuation practices. On the one hand, IPOs entail general, institutional investors who do not know how to value life sciences R&D; and on the other hand, trade sales involve specialist purchasers (e.g. other life sciences firms) who have specific knowledge about the value of the R&D to them or other firms (Birch forthcoming). Consequently, as one informant pointed out:

> ... there's a different valuation metrics in terms of the way industry [i.e. trade sale] will value business and investor [i.e. IPO]. So you get a materially higher premium or exit value if you sell than if you trade the shares.
>
> (Broker H)

R&D strategies

Whether or not financing strategies, processes, ecosystems, and actors have changed may seem tangential to whether research and innovation strategies have shifted. Are life sciences firms still researching and developing groundbreaking therapies, for example? This is a difficult question to answer, as one informant noted even though they agreed it is likely to have happened (BA Syndicate Investor L). Existing literature suggests that financialization has, for example, reconfigured R&D in pursuit of IP that can be capitalized (Pisano 2006); has diverted funds from R&D to share buybacks in order to boost share prices (Lazonick and Tulum 2011); and has created problematic pressures on decisions about disease targets, orphan drug status, and such like (Birch 2006; Andersson *et al.* 2010; Birch *et al.* 2016). One specific example of these changes that came out of the interviews revolved around the increasing involvement of large pharmaceutical companies (and other institutions like the Wellcome Trust) in *strategic investing* (e.g. VCT Investor E). Basically, pharmaceutical firms like GlaxoSmithKline and AstraZeneca have expanded their venture investing operations since the GFC with venture funds like SR One and MedImmune Ventures respectively. The rationale behind this form of financing, as opposed to acquisitions or direct spending, is that it outsources R&D risks to small life sciences firms, something I have described elsewhere as the 'governance of risk outsourcing' – see Chapter 4 (Birch and Cumbers 2010). Informants were quite explicit about this strategy, explaining that 'a lot of pharma's, for instance, are effectively outsourcing their R&D' (Venture Debt Investor D). Even clearer is another who said that:

> ... the reason they [pharmaceutical firms] have venture arms is they want, in some way, not to take on too much early stage risk into their pipeline ... so if they have to write off an investment in a few million quid in some daft company that doesn't work, they don't take that hit in public markets, so they sort of have these corporate venture arms to create sort of R&D without any risk.
>
> (BA Syndicate Investor L)

After looking at these five issues it is worth considering whether the GFC did *actually* have a significant impact on financing in the life sciences. Despite all the points raised above, it is interesting to note that many informants, when asked whether the GFC had had an impact or not, claimed there was no real difference between before and after the crisis (at time of speaking in 2012). Take this informant, for example, who said that:

> I don't think myself that there's any significant difference between ... before and after crisis and post ... I don't think there's any fundamental change in venture capital activity, other than a sort of general tightening of the market.
> (VC Investor K)

Other informants adhered to this view (e.g. VC Specialist Investor C). My take on this is that the financing *process* has not changed, rather than that the GFC had no discernible impact. To this informant, for example, the chain or relay of finance I highlighted above in the analytical discussion has remained, largely, the same (see Andersson *et al.* 2010; Hopkins 2012; Birch forthcoming). This reflects a change in circumstance (i.e. amount of capital available) rather than substance (i.e. changing financing strategies).

Conclusion

In this chapter I have discussed the financing of the life sciences by drawing on a range of theoretical literature and empirical material. My starting point is that the life sciences sector has been reconfigured by the rise of finance and financial markets over the last 30–40 years, just like other sectors of the economy (see Krippner 2005; Pike and Pollard 2010; Coe *et al.* 2014). It is important, therefore, to think about how the life sciences sector has been financialized (Andersson *et al.* 2010), and what this means for pursuit of particular business and research strategies. It is also important, in light of global events over the last few years, to analyze how the global financial crisis (GFC) has impacted on the life sciences.

As the empirical material I marshal in this chapter illustrates, the life sciences has been significantly affected by the GFC. It has reshaped life sciences financing, with the decline of venture capital and IPOs and rise of debt and partnering income, as well as the particular business and research strategies pursued by individual firms, financiers and investors. In particular, the GFC has led to a wholesale shift in the UK life sciences towards shorter-term strategies based on trade sales, which is underpinned by the outsourcing of R&D by large pharmaceutical firms. In the next chapter I consider the implications of these findings to broader issues around regional development.

Note

1 This argument by Mirowski (2012) is supported by the comment by one informant that 'it is a bit of, can be a bit of a Ponzi scheme which is, you know, [phone rings], "I pay this for it, so that's what it's worth"'(C VC Specialist Investor 2012).

References

Andersson, T., Gleadle, P., Haslam, C. and Tsitsianis, N. (2010) Bio-pharma: A financialized business model, *Critical Perspectives in Accounting* 21: 631–641.

Birch, K. (2006) The neoliberal underpinnings of the bioeconomy: The ideological discourses and practices of economic competitiveness, *Genomics, Society and Policy* 2(3): 1–15.

Birch, K. (2007) The virtual bioeconomy: The 'failure' of performativity and the implications for bioeconomics, *Distinktion: Scandinavian Journal of Social Theory* 8(1): 83–99.

Birch, K. (2008) Alliance-driven governance: Applying a global commodity chains approach to the UK biotechnology industry, *Economic Geography* 84(1): 83–103.

Birch, K. (2011) 'Weakness' as 'strength' in the Scottish life sciences: Institutional embedding of knowledge-based commodity chains in a less-favoured region, *Growth and Change* 42(1): 71–96.

Birch, K. (2012) Knowledge, place and power: Geographies of value in the bioeconomy, *New Genetics and Society* 31(2): 183–201.

Birch, K. (2015) *We Have Never Been Neoliberal*, Winchester: Zero Books.

Birch, K. (forthcoming) Rethinking *value* in the bio-economy: Finance, assetization and the management of value, *Science, Technology and Human Values*.

Birch, K. and Cumbers, A. (2010) Knowledge, space and economic governance: The implications of knowledge-based commodity chains for less-favoured regions, *Environment and Planning A* 42(11): 2581–2601.

Birch, K. and Mykhnenko, V. (2014) Lisbonizing vs. financializing Europe? The Lisbon Strategy and the (un-)making of the European knowledge-based economy, *Environment and Planning C* 32(1): 108–128.

Birch, K. and Tyfield, D. (2013) Theorizing the bioeconomy: Biovalue, biocapital, bioeconomics or ... what?, *Science, Technology and Human Values* 38(3): 299–327.

Birch, K., Levidow, L. and Papaioannou, T. (2010) *Sustainable Capital?* The neoliberalization of nature and knowledge in the European knowledge-based bio-economy, *Sustainability* 2(9): 2898–2918.

Birch, K., Levidow, L. and Papaioannou, T. (2014) Self-fulfilling prophecies of the European knowledge-based bio-economy: The discursive shaping of institutional and policy frameworks in the bio-pharmaceuticals sector, *Journal of the Knowledge Economy* 5(1): 1–18.

Birch, K., Dove, E., Chiappetta, M. and Gursoy, U. (2016) Biobanks in oral health: Promises and implications of 'post-neoliberal' patterns of science and innovation, *OMICS: A Journal of Integrative Biology* 20(1): 36–41.

Boyer, R. (2000) Is a finance-led growth regime a viable alternative to Fordism? A preliminary analysis, *Economy and Society* 29(1): 111–145.

CEC (2010) *Communication from The Commission. Europe 2020: A Strategy for Smart, Sustainable and Inclusive Growth* COM(2010) 2020 final, Brussels: Commission of the European Communities.

Cleeton, D. (2012) Evaluating the performance record under the Lisbon agenda, in M. Smith (ed.) *Europe and National Economic Transformation: The EU after the Lisbon Decade*, Basingstoke: Palgrave Macmillan, pp. 15–31.

Coe, N., Lai, K. and Wojcik, D. (2014) Integrating finance into global production networks, *Regional Studies* 48(5): 761–777.

Cooke, P. (2007) *Growth Culture*, London: Routledge.

Cooper, M. (2008) *Life as Surplus*, Seattle: University of Washington Press.

Deeg, R. (2012) Innovation financing in Europe: What has financial market integration brought?, M. Smith (ed.) *Europe and National Economic Transformation: The EU after the Lisbon Decade*, Basingstoke: Palgrave Macmillan, pp. 74–94.

European Council (2000) *An Agenda of Economic and Social Renewal for Europe* (aka Lisbon Agenda), Brussels: European Council [DOC/00/7].

EY (2015) *Biotechnology Industry Report 2015: Beyond Borders*, Boston, MA: EY LLP.

Grahl, J. (2011) The subordination of European finance, *Competition and Change* 15(1): 31–49.

Hopkins, M. (2012) Exploring funding routes for therapeutic firms, in M. O'Neill and M. Hopkins (eds) *A Biotech Manager's Handbook*, Oxford: Woodhead Publishing.

Hopkins, M., Crane, P., Nightingale, P. and Baden-Fuller, C. (2013) Buying big into biotech: Scale, financing, and the industrial dynamics of UK biotech, 1980–2009, *Industrial and Corporate Change* 22(4): 903–952.

Huggett, B. (2013) Public biotech 2009 – the numbers, *Nature Biotechnology* 31: 697–703.

Huggett, B. and Lähteenmäki, R. (2012) Public biotech 2011 – the numbers, *Nature Biotechnology* 30: 751–757.

Huggett, B., Hodgson, J. and Lähteenmäki, R. (2009) Public biotech 2008 – the numbers, *Nature Biotechnology* 27: 710–721.

Huggett, B., Hodgson, J. and Lähteenmäki, R. (2010) Public biotech 2009 – the numbers, *Nature Biotechnology* 28: 793–799.

Huggett, B., Hodgson, J. and Lähteenmäki, R. (2011) Public biotech 2010 – the numbers, *Nature Biotechnology* 29: 585–591.

Krippner, G. (2005) The financialization of the American economy, *Socio-Economic Review* 3: 173–208.

Lähteenmäki, R. and Lawrence, S. (2005) Public biotechnology 2004 – the numbers, *Nature Biotechnology* 23: 663–671.

Lawrence, S. and Lähteenmäki, R. (2008) Public biotech 2007 – the numbers, *Nature Biotechnology* 26: 753–762.

Lawrence, S. and Lähteenmäki, R. (2014) Public biotech 2013 – the numbers, *Nature Biotechnology* 32: 626–632.

Lazonick, W. and O'Sullivan, M. (2000) Maximizing shareholder value: A new ideology for corporate governance, *Economy and Society* 29(1): 13–35.

Lazonick, W. and Tulum, O. (2011) US biopharmaceutical finance and the sustainability of the biotech business model, *Research Policy* 40(9): 1170–1187.

Leyshon, A. and Thrift, N. (2007) The capitalization of almost everything: The future of finance and capitalism, *Theory, Culture and Society* 24(7–8): 97–115.

Mazzucato, M. (2013) *The Entrepreneurial State*, London: Anthem Press.

Miller, D. (2002) Turning Callon the right way up, *Economy and Society* 31(2): 218–233.

Mirowski, P. (2012) The modern commercialization of science as a Passel of Ponzi schemes, *Social Epistemology* 26(3–4): 285–310.

Mittra, J. (2016) *The New Health Bioeconomy*, Basingstoke: Palgrave Macmillan.

Morrison, C. and Lähteenmäki, R. (2015) Public biotech 2014 – the numbers, *Nature Biotechnology* 33: 703–709.

OLS (2011) *Strategy for UK Life Sciences*, London: Department of Business, Innovation and Skills.

Pike, A. and Pollard, J. (2010) Economic geographies of financialization, *Economic Geography* 86(1): 29–51.

Pisano, G. 2006. *Science Business*, Boston, MA: Harvard University Press.

Ponte, S. (2009) From fishery to fork: Food safety and sustainability in the 'virtual' knowledge-based bio-economy (KBBE), *Science as Culture* 18(4): 483–495.

Powell, W., Koput, K., Bowie, J. and Smith-Doerr, L. (2002) The spatial clustering of science and capital: Accounting for biotech firm–venture capital relationships, *Regional Studies* 36(3): 291–305.

Rosamond, B. (2002) Imagining the European economy: 'Competitiveness' and the social construction of 'Europe' as an economic space, *New Political Economy* 7(2): 157–177.

Scannell, J. (2015) *Four Reasons Drugs are Expensive, of Which Two are False: An Opinion*. University of Edinburgh: Innogen Working Paper No.114.

Sokol, M. (2015) Financialisation, financial chains and uneven geographical development: Towards a research agenda, *Research in International Business and Finance*, doi:10.1016/j.ribaf.2015.11.007.

Styhre, A. (2014) Coping with the financiers: Attracting venture capital investors and end-users in the biomaterials industry, *Technology Analysis and Strategic Management* 26(7): 797–809.

Styhre, A. (2015) *Financing Life Science Innovation*, Basingstoke: Palgrave Macmillan.

Sunder Rajan, K. (2006) *Biocapital*, Durham: Duke University Press.

Sunder Rajan, K. (ed.) (2012) *Lively Capital*, Durham: Duke University Press.

Waldby, C. and Mitchell, R. (2006) *Tissue Economies: Blood, Organs and Cell Lines in Late Capitalism*, Durham: Duke University Press.

Yang, W. (2014) 2013 – biotech back in the saddle, *Nature Biotechnology* 32(2): 126.

7 Conclusion

Innovation and regional development for whom?

Introduction

The purpose of this book has been to examine the relationship between innovation, regional development and the life sciences. Up to now, much of the book has focused on innovation and the life sciences. I deliberately left the concerted discussion of regional development until this chapter, meaning that it represents a concluding discussion for the whole book. I wrote it this way so that I could focus on the topic of regional development in a standalone chapter, bringing all the research to bear on this particular issue rather than spread it out. Consequently, my intention in this chapter is to focus on the implications for regional development of my findings about the life sciences sector and innovation in the previous chapters. Moreover, rather than write a separate concluding chapter, I thought it best to finish the book here.

As a starting point for this chapter, then, it is important to have a clear stance on what I mean by regional development. As Pike *et al.* (2006) note in their book, *Local and Regional Development*, the term 'regional development' covers a range of concerns and issues that need unpacking. First, regional development requires a discussion of *what kind* of development we are talking about; for example, is it 'economic', is it 'socio-economic', is it 'sustainable', and so on? As Pike *et al.* (2006) note, much of the literature on regional development equates economic concerns with 'development' (e.g. more jobs, more firms, more income, more taxes, etc.), setting aside considerations of the social dimensions of regional development (e.g. social cohesion, enabling opportunities, household equity, etc.). Second, regional development necessarily entails normative questions about *for whom* it is being undertaken; for example, is it for employers, is it for workers, is it for local government, and so on? Again, Pike *et al.* (2006) point out that regional development should entail regional decision-making about the social-economic priorities that people want for their localities. In particular, it necessitates an acknowledgement that one-size does not fit all when it comes to policy-making (see Lovering 1999). In light of these arguments, I define regional development in socio-economic terms as the improvement of people's economic, social and political lives as they seek to make and remake the places in which they live.

Innovation plays a major role in regional development, whatever definition we use. As the literature I discuss in Chapter 2 – as well as throughout the book – illustrates, innovation is characterized as the key driver behind the emergence of new products, services, and whole industries, enabling the replacement of old, stagnating sectors in response to increasingly influential global imperatives like competitiveness. The life sciences, in this regard, are frequently touted as an innovative sector *par excellence*, representing the promise of high value and high-tech development for all regions. Whether or not the life sciences holds such promise is addressed in this chapter. Despite the valorization of innovation as *the* driver of regional development, it is also worth noting some of the costs associated with new, high-tech sectors. As Walker (2006) has pointed out, regions like Silicon Valley may experience unprecedented growth during economic booms, but they also experience huge decline when there is a crash (e.g. the 2000 DotCom Crash). The dependence of such regions on high-tech sectors, which are underpinned by volatile financing, flexible labour markets and uncertainty, necessarily involves significant economic social and political dislocation for regional residents as they lose work, incomes, livelihoods, and so on. It is important, in this sense, to avoid a rose-tinted view of innovation when it comes to regional development, especially in relation to high-tech sectors like the life sciences.

In the rest of this chapter I return to the findings from each empirical chapter. I do so in order to consider the following issues: what development is happening? For whom is it happening? And what are the implications for life sciences innovation as a result? I cover the geographies of innovation; innovation governance; innovation imaginaries; and innovation financing. I discuss each in turn and highlight how the findings problematize simplistic assumptions that the life sciences sector – or innovation more generally – represents a panacea to regional socio-economic decline resulting from processes of deindustrialization, globalization and financialization. Rather, I consider how the life sciences have come to represent a mythic or fetishized sector in debates around regional development, often being presented as a key example of and policy for regional renewal.

Regional development and innovation geographies

During the late 1990s and early 2000s, the UK 'New' Labour government led a resurgence of regional policy initiatives after the erosion of regional policy during the preceding Conservative administrations, especially during the Thatcher years (Birch *et al.* 2010a). The two key examples of the re-emergence of the 'region' in government policy were (1) the devolution of power to national assemblies in Scotland, Wales and Northern Ireland, and (2) the establishment of Regional Development Agencies (RDAs) across England in 1998.

While the UK Labour government pursued a relatively strong regional policy, it was underpinned by the concept of regional and national *competitiveness* in which economic growth – as opposed to development – was premised on rising productivity, especially resulting from the expansion of the 'knowledge economy' (e.g. DTI 1998). In this policy context, the life sciences sector came to represent an iconic exemplar of the knowledge economy and a certain form of regional development, namely one driven by support for localized, industrial 'clusters' (e.g. DTI 1999a, 1999b).

Wither the knowledge economy? As subsequent research has shown (e.g. Birch and Myhnenko 2009, 2014), the focus on knowledge, knowledge-based sectors and the knowledge economy perhaps hides more than it illuminates in relation to regional development. Growth in high-tech manufacturing and services, for example, has been highly uneven across British (and European) regions. In many cases, it would be hard to claim that there has been a wholesale transition to a knowledge economy in many or most UK regions, especially in the more disadvantaged regions of the country (e.g. Midlands and Northern England). In this book, my research illustrates the extent to which the life sciences were – and still are – concentrated in particular regions, including Oxfordshire, Cambridgeshire, London and Eastern Scotland. Even within these regions, there was limited evidence that the life sciences sector represented a major employer or major contributor to local economic growth. As a result, this led me to the conclusion that the life sciences sector did not (and cannot) represent a viable development alternative to other sectors (e.g. services, manufacturing) when it comes to increasing regional employment, incomes, and productivity. The fact that there were only around 450 life sciences firms across the whole of the UK illustrates the extent to which it pales in importance against other sectors when it comes to promoting regional employment, for example, whether or not it produces new innovative technologies.

Localized knowledge, learning and innovation: the 1990s and early 2000s were a period of intense debate about regional (economic) dynamics, across disciplinary fields. Much of this literature stressed the localized, systemic and collective nature of innovation, largely based on notions of localized knowledge spillovers and human capital mobility (e.g. Audretsch and Stephan 1996; Zucker *et al.* 1998; Cooke 2007). A literature stressing the importance of extra-local interactions and linkages began to emerge in response (e.g. Bathelt *et al.* 2004; Coenen *et al.* 2004; Leibovitz 2004; Zeller 2004; Gertler and Levitte 2005). I discuss my contribution to this debate in the next section, but for now it is useful to highlight my empirical findings in Chapter 3. This data, drawn from my PhD research, showed that even supposedly 'clustered' life sciences firms drew more on national and international codified *and* tacit knowledge than local knowledge – even local tacit knowledge (cf. Gertler 2003). As such this research supported the more sceptical views of Malmberg (2003) and Malmberg and Power (2005) on the extent and depth of interactions, spillovers, etc. in regional economies.

Consequently, it problematized the idea that knowledge, learning and innovation are regionally-centred or -based; instead, it is more apt to think of them as cross-scalar or multi-scalar.

Multi-scalar processes: it became obvious that the distinction between local, national and global scales in understanding and supporting regional development was rather limiting, analytically and policy-wise. As Phelps (2004) has noted, there is a tendency in the regional studies literature to link the 'regional economy' with the 'global economy', situating the one in relation to the other (e.g. Amin and Thrift 1992; Bathelt *et al.* 2004). This 'binary distinction', in Phelps' (2004) words, obscures the overlapping, entangled and co-produced nature of geographical processes when it comes to innovation and regional development. As my own research shows, the life sciences sector is embedded in a set of *multi-scalar* processes, it is not intrinsically localized or regional (Birch 2009). For example, the knowledge–space dynamic in the UK life sciences entails: (a) interaction of local, national and global knowledge exchange between firms and universities; (b) linkages between local, national and global business, research and regulatory organizations; and (c) spillovers between local, national and global social actors.

One-size-fits-all policy: perhaps the clearest finding that my research in Chapter 3 provides is that regional economies are distinct from one another; for example, the four UK life sciences concentrations I identified had very different *knowledge–space dynamics*. As a truism, this claim is rather banal. However, it is important to remember when considering policy-making decisions and effects. The emphasis on competitiveness that underpinned UK regional policy was enacted through support for industrial clusters (Porter 1990, 2000), establishing 'clusters' as a suitable instrument for regional development across regional economies irrespective of the starting conditions of each region. For example, the government introduced an *Innovative Clusters Fund* in 2000, to finance incubators and cluster infrastructure, and a *Regional Innovation Fund* in 2001 (DTI 2003). Other policies included the introduction of *Business Planning Zones* to support 'flexible planning' for establishing clusters in deprived areas (HM Treasury 2003). In relation to the life sciences, this one-size-fits-all tendency was evident in the fact that all the English RDAs identified 'biotechnology' or 'life sciences' as one of the key sectors they sought to support, regardless of the extent of the sector in the region and the country more generally. Each region, in this sense, must find its own way when it comes to regional policy-making (Gertler and Vinodrai 2009).

The implications of these findings to regional policy-making are significant. As mentioned already, the very idea of a one-size-fits-all policy does serious injustice to the existing and ongoing uneven development of regional economies in the UK. It makes no sense, for example, to promote the life sciences sector as the answer to the decline in primary or manufacturing industries resulting from the wholesale (neo-liberal) restructuring of these industries during the Thatcher administration of the 1980s (see Birch *et al.* 2010a). Leaving aside the impacts of that restructuring (e.g. long-term unemployment, poor health, social dislocation, etc.), it is not credible that a 'less-favoured' regional economy can (or should) base its renewal on a high-tech sector that is dependent on a limited number of highly-educated scientists and researchers. This leaves us with key questions like, where is everyone else going to work? And what conditions of work are available to them? How will regional economies position themselves in knowledge-based sectors? And so on.

Regional development and innovation governance

Less-favoured regions (LFRs) face particular difficulties in the knowledge economy, especially in terms of how and where they position and reposition themselves in global commodity or value chains (GC/VC) (Birch and Cumbers 2010; Birch 2011). I address these issues in Chapter 4 by conceptualizing the life sciences sector as a *knowledge-based commodity chain.* Although GC/VC research has tended to focus on issues of technological and organizational upgrading in developing countries resulting from these value chains (e.g. Gereffi 1994; Bair 2005; Gereffi *et al.* 2005), it is not clear whether this is also possible when it comes to regional development within a (developed) country. There is a real threat that social actors in growth regions simply sideline LFRs, positioning themselves at key junctions in the value chain to the exclusion of other regions. Any regional restructuring in pursuit of this notional knowledge economy, therefore, needs to consider very carefully the socio-economic implications of that restructuring. Several issues are pertinent in this regard: it is not clear that the knowledge economy will alleviate uneven regional development; it is not clear whether the 'success' of the knowledge economy in some regions (e.g. Silicon Valley) will hinder this uneven regional development; and it is not clear how the power differences between growth and less-favoured regions will reinforce existing political-economic arrangements and flows of value, resources and workers between regions.

Alliance-driven governance: as I have tried to illustrate across the chapters in this book and others have discussed elsewhere (e.g. Godin 2006), innovation is complex and belies the easy conceit of conceptualizing it as a linear process. Due to the complexity of new products and services, innovation in knowledge-based commodity chains – like the life sciences – entails the coordination of an array of often highly diverse and geographically dispersed economic actors (e.g. universities, businesses, governments, regulators, etc.). As such, innovation governance has to be alliance-driven (Birch 2008); it involves bridging the economic, social and political differences between these diverse and dispersed actors in ways to encourage new forms of collaborative working. Regional development policies have to recognize that single firms cannot do it all themselves; in this sense, the idea of 'open innovation' pioneered by Chesbrough (2003) offers an opportunity for regional policy-makers. First, it promotes the idea of collaboration – whether locally or extra-locally – and, second, such collaboration can provide an opportunity for regions to attract new resources and capabilities from elsewhere. The downside to this, though, is that collaboration could turn firms into more attractive acquisition targets, with the risk of losing value created in places like Scotland to other regions and countries.

Positioning along commodity chains: in light of the need to engage in new forms of innovation governance, it is important to understand how firms and regions can position themselves within global commodity or value chains. Dominant markets, like the USA, can have a significant impact on LFRs like Scotland when it comes to regional development. Sectors like the life sciences, for example, are necessarily 'born global' as one informant told me. Firms have to compete globally and orient themselves globally to engage in knowledge exchange and upgrading along knowledge-based commodity chains. As a result, many LFRs are more reliant on extra-regional linkages and relationships than other regions, especially when firms are nearing the market (i.e. they have a product for sale). For example, one particular area where Scottish life sciences firms were dependent on external capabilities was the need to attract managers and directors to Scotland. The global nature of the life sciences, especially centred on the USA, meant that such individuals were rare in Scotland, having been attracted to other places where there were more employment opportunities. Attracting them back to Scotland was sometimes difficult, meaning that regional development policies have to do more than anchor high-tech sectors through public spending, they also have to tie regions into global networks (e.g. labour markets).

Less-favoured regions in the knowledge economy: as the findings in Chapter 4 show, innovation and the knowledge economy do not alleviate uneven development themselves. The life sciences are concentrated in particular regions, especially growth regions like South-East and Eastern England (Birch 2009). Although Scotland represents another concentration, the position of Scottish life sciences firms at certain points in the knowledge-based commodity chain illustrates the extent to which regional development is driven by extra-local processes. For example, the lack of a market for life sciences products and services in Scotland meant it was hard to keep firms grounded as well as growing. Consequently, most firms needed to look elsewhere at a certain point in their evolution; either for partners, or buyers, or investors. This has implications for regional development, since it is not possible to copy the 'successful' research or business strategies pursued by US firms – international 'best practice' or 'benchmarking' might end up harming regional development policy, in this sense, rather than providing useful insights for regional policy-makers. Other strategies are necessary, and regional policies to support those strategies are necessary. This might mean taking a more modest policy approach in which the aim is only to develop and support firms up to the point that they become viable acquisition targets, with the secondary aim that the capital returned from such sales ends up recycled back into the regional economy (Birch 2011).

With these findings in mind, it is worth considering how LFRs might position themselves in knowledge-based commodity chains that are, necessarily, global and coordinated through alliance-driven forms of governance. As already pointed out, this might require that policy-makers develop more modest regional policy goals; for example, support firms up to a certain point and try to capture the capital from their sale. It also requires that policy-makers have a clear idea of the local or regional institutional context, reflecting the debate on 'new regionalism' in the literature (e.g. Amin and Thrift 1992; Morgan 1997; Cooke and Morgan 1998; Lovering 1999; Pike *et al.* 2006). However, rather than simply focus on a region's institutional *strengths* – in the case of the Scottish life sciences, this would include international universities – it is also important to understand how supposed institutional *weaknesses* are potential strengths. For example, my research on the Scottish life sciences showed that the population size of Scotland was a major stumbling block for firms trying to introduce new products and services, because the (regional) market for them was so small. Although a major drawback on the one hand, the size also provided certain benefits; namely, it increased trust because social actors knew each other and could not exploit each other without consequences to their reputation (Birch and Cumbers 2010; Birch 2011). Obviously, there is a downside to this as well, but if regional policy-makers can develop policies that translate these weaknesses into an advantage, then it is more likely to benefit the regional economy than the pursuit of the next Amgen.

Regional development and innovation imaginaries

My focus in Chapter 5 was the future visions, promises, expectations and so on that surround life sciences innovation everywhere. Such *discourse* is important for regional development and regional policy as much as it is for research and innovation policy. As a number of academics have argued, discourses are performative – by which they mean, they have material effects reflecting the discursive claims (e.g. Callon 1998; Loconto 2010). Others maintain that while discourse has an effect, it might not be 'performative' in the sense that the effect does not reflect the discursive claims (e.g. Birch 2007; Ponte 2009). I tend towards the latter view, which means I prefer to use the concept of *imaginaries* to understand how these future visions, promises, expectations, etc. impact on regional development. As discussed in Chapter 5, imaginaries are ways to enrol people and resources, as well as shape and reshape institutions and policy frameworks (Birch *et al.* 2010b, 2014). In the European context, I showed how imaginaries describe the world in certain ways, how these descriptions lead to normative choices, and how these choices engender institutional and policy changes to enact the imaginaries. They are 'impactful', in this sense, and in a number of ways:

Imagining biotech clusters: in his well-known book, *The Competitive Advantage of Nations*, the management professor Michael Porter (1990) argued that competitiveness was based on (national) industrial cluster (e.g. Italian ceramics, Japanese robotics, etc.). In his later work, Porter (2000) defined clusters in geographical terms as localized clusters of firms and interdependent institutions. This concept was seized upon by UK policy-makers, who flipped the claims on their head to assume that clustering engendered innovation and competitiveness. Examples of this in the life sciences include the DTI's *Biotech Clusters* and *Genome Valley* reports (DTI 1999a, 1999b). As a result of the cluster concept and its embedding in policy circles, there have been numerous attempts by regional policy-makers to 'create' regional clusters, especially life sciences clusters (e.g. OBN 2002; Oxford Trust 2002). Such cluster imaginaries obscure the distinctiveness of each regional economy, which I outline in Chapter 3, and present certain regional qualities, regional linkages and regional institutions as universal prerequisites for regional development. There is limited evidence to support such locally-bounded approaches (Phelps 2004; Malmberg and Power 2005), but this has not stopped the formation of life sciences clusters. Such clusters, in the UK at least, provide their members with an identity that helps attract financing, workers, etc., even though firms must, necessarily, connect into global networks.

Reconfiguring policy frameworks: much of Chapter 5 focuses on the influence of imaginaries in driving institutional and policy changes in European Union innovation circles. As the chapter shows, innovation imaginaries entail descriptive, normative and performative elements; for example, one powerful discursive trope goes something like, 'Europe lags behind the USA because it doesn't spend enough on R&D; therefore Europe needs to support institutional and policy changes that increase R&D spending'. Through such institutional and policy change, however, private firms can gain access to public research funds, public-sector capacities, common natural resources, product approvals, intellectual property, etc. As a result of enacting these innovation imaginaries, institutional and policy changes end up blurring the boundary between commercial interests and societal needs, as well as between the private and public sectors. These would-be solutions exemplify 'the competition state', whereby the state promotes domestic capacities for global economic competition (cf. Cerny 1999) – even if competitors are increasingly trans-national companies, whose global integration is further helped by policies for 'competitiveness' (van Apeldoorn, 2000). When this comes to regional development, there is a real possibility that innovation imaginaries, especially in the fetishizing of innovation as the source of socio-economic development, end up enrolling regional policy and regional resources in the pursuit of transnational, corporate interests rather than local and regional needs and priorities.

Geographical imaginaries: shortly after the Conservative-Liberal Democract Coalition government came to power in the UK in 2010, it released a policy document, *Local Growth: Realising Every Place's Potential* (BIS 2010), which stated that the 'region' is an artificial concept: 'Not only did this approach [reducing growth rates between regions] lead to policies which worked against the market, it was also based on regions, an artificial representation of functional economies' (p. 7). Primarily a justification for abolishing the Regional

Development Agencies, this example illustrates the extent to which geographical locations are imaginary; that is, geographies are imagined as having certain qualities, or representativeness, or problems, and so on. When it comes to the life sciences, this reflects the imagining of opportunities and problems facing the sector, especially through the comparison of geographically-defined jurisdictions (e.g. countries and regions) with one another. As the findings in Chapter 5 show, the USA is held up as the example to emulate, despite the fact that no other jurisdiction (e.g. EU) has the same institutional or policy conditions. This is used, though, to define the areas that other jurisdictions have to change to be more like the USA. In regional development terms, this frequently leads to a one-size-fits-all policy narrative in which regional policy is driven by unrealistic expectations (Gertler and Vinodrai 2009).

In the regional policy context, the life sciences sector represents an iconic exemplar of the knowledge economy and a certain form of regional development, namely one driven by support for localized, industrial 'clusters' (e.g. DTI 1999a, 1999b). Support for this form of regional development can cut across government departments; for example, in the UK it involves not only the DTI (2003), but also DETR (2000), HM Treasury (2001, 2003), and ODPM (2004). It is important to emphasize that this form of regional development is not without its academic support, building on debates about the collective, systemic and social processes that underpin regional growth, innovation and clustering (for reviews, see Asheim and Gertler 2005; Fagerberg 2005; for critique, see Malmberg and Power 2005). However, there is a certain conceptual slippage between the academic literature (i.e. discourse) and policy implementation (i.e. enactment) in which empirical findings (e.g. 'clustered firms are more innovative') are turned on their head (e.g. 'innovation results from clustering'). My research discussed in Chapter 5 illustrates how discourses come to shape institutional and policy frameworks in ways that confirm the original discursive claims. With regional policy, there is a real threat that innovation imaginaries come to reshape regional institutions and policy-making in ways that reflect a mythic and universal image of the life sciences sector, rather than a geographically-specific sector grounded in regional institutions and contexts. The end result is the fetishization of innovation in regional policy, such that 'non-innovative' sectors (e.g. services) are ignored, devalued, and degraded threatening the overall regional economy.

Regional development and innovation financing

The final empirical chapter in the book focused on financing innovation. As I note, the literature has generally focused on the co-location of the life sciences and venture capitalists (e.g. Powell *et al.* 2002), rather than on practices and processes involved in understanding innovation financing. It is important, as Coe *et al.* (2014) point out, to understand how finance impacts on regional economies and development since the financial process – or financialization – is, by its nature, porous, stretching across borders. At the same time, Coe *et al.* argue that we have to understand how firms are financed in order to properly theorize global value chains or production networks. Moreover, though, it is important to understand how financialization affects the actions, decisions and strategies of life sciences firms and regional policy-makers, since it changes how people understand things like value and valuation (Chiapello 2015). In my research I identity three ways that these issues might be important:

Financialization and innovation: much of the research I reported on in Chapter 6 demonstrates the extent to which finance and financial value have come to underlay thinking in the life sciences sector specifically and other innovative sectors more generally. As Eve Chiapello (2015) argues, the way we conceptualize value impacts on the social, political and normative judgements we make; for example, financial valuations – such as discounted notions of the 'time value' of money – come to frame our thinking (e.g. in terms of opportunity costs). Evidence of how such financial thinking frames life sciences innovation is evident in the differences between business and research strategies before and after the global financial crisis (GFC). After the GFC there was more emphasis on trade sales as opposed to accessing capital through an initial public offering (IPO) in order to push forward product development. Other changes included the expansion of corporate venturing by large pharmaceutical firms in order to outsource the risk represented by research failures to shareholder value. As a number of scholars note, technoscientific research and innovation are bound up with the financing of these endeavours, they are not distinct or neutral factors (Mirowski 2011; Tyfield 2012a, 2012b).

Innovation and financial ecosystems: as I note above, life sciences innovation is shaped by the financial ecosystem in which it happens. Others have made similar claims (e.g. Mittra 2016; Owen and Hopkins 2016). However, this insight is often framed in terms of the need to adapt to the US model of high-tech development, in which the emergence of new sectors is underpinned by loosely-regulated and risk-friendly venture capital (VC) and capital markets. Institutionally-speaking, this is problematic. As other research shows (e.g. Birch and Mykhnenko 2014), knowledge-based sectors like the life sciences do not simply emerge from deregulated financial markets, the reverse could be the case. Recent work by Mariana Mazzucato (2013) highlights the central role of the state and public sector in providing the early-stage funding, and long-term support, needed for new sectors to emerge and develop. As such, the financial ecosystem is important for regional development, but it is as important to ensure that the ecosystem is suited to regional economies; most are not, however, because they are – as mentioned – not restrained by geographical boundaries (Coe *et al.* 2014). It makes little sense, then, for most regions to push their firms, life sciences or otherwise, towards public market IPOs because most of the value in those firms will simply be lost. Moreover, those firms' strategies will end up driven by priorities set outside the region, predominantly in major financial centres like London, New York and Toronto.

Finance and uneven regional development: one crucial aspect of high-tech financing is the concentration of VC in particular regions (Powell *et al.* 2002). In the UK, for example, most VC is centred on London and, to a lesser extent, the growth regions of Cambridgeshire and Oxfordshire. As a result, the life sciences sector outside of these regions faces difficulties in accessing capital for investment. In some places like Scotland this unevenness has led to the emergence of alternative sources or processes of financing, including strong business angel syndicates (Birch 2011). While these alternatives may offer certain advantages – such as recycling capital back into the region when a firm is sold – the unevenness of VC financing places limits on what policy-makers can do to encourage regional investment. When it came to Scotland, for example, a critical financing gap was the lack of 'lead investors' in Scotland; these were people who would bring other investors on board. Other places, like London, had such lead investors, but the focus of VC on their locality meant that they rarely looked elsewhere for opportunities. As a result, Scottish firms end up limited to a certain size before they are (generally) sold to larger businesses. In the end, this makes it difficult for a large, anchoring firm to emerge in LFRs.

Conclusion

I want to finish this book by considering the *problem of innovation* in regional development. Since the early 2000s, there has been a growing fear about a supposed 'productivity crisis' in the biopharmaceutical industry. In particular, a number of academics, commentators, policy-makers, business people, and so on have sought to frame the relationship between R&D investment in biopharmaceuticals and the emergence of new products as increasingly problematic because the number of new products has not increased alongside increasing R&D spending (Pisano 2006; Hopkins *et al.* 2007). More recently, this fear has transformed into a concern about healthcare remuneration policies, especially in the Global North, impacting negatively on biopharmaceutical R&D (Ernst and Young 2015). Such fears have always, it is important to note, accompanied the emergence and development of the life sciences sector. The UK's 1980 *Spinks Report*, for example, would not seem that out of place in today's policy debates, centred, as it is, on the promises of 'biotechnology' for solving society's problems. However, the life sciences sector has simply not lived up to expectations; or, at least, not up to the research and innovation expectations of new products and services to resolve multiple, overlapping societal challenges (e.g. old age, inequality, sustainability, disease, etc.).

Rather than a technical problem that can be solved with more investment in R&D, this failure to meet expectations results from the peculiar characteristics of the life sciences sector. For example, while the life sciences sector is frequently presented or fetishized as a dynamic, entrepreneurial, innovative, high-tech and high-risk sector, it is dominated more by IP rights, regulated monopolies (e.g. orphan drug designation) and market exclusivity. As a result, product development is often driven as much by political-economic concerns (e.g. commercial profit) as by technical ones (Birch 2006). Innovation, in this regard, has become a process to delivering monopoly super- or quasi-rents (Zeller 2008), predominantly through the growth of IP. Unfortunately, this can have a negative impact on research and innovation, yet it does not necessarily have negative implications for the growth and expansion of the life sciences sector in financial terms – the opposite in fact. According to Ernst and Young (2015), the global biotech industry passed US$1 trillion in market capitalization in 2014 and did so *because*, and not despite, the fact it is a highly financialized sector. Such share valuations are secured through financial practices for valuing IP assets and through the active intervention of firms in shoring up their share value through buy-backs and suchlike (Andersson *et al.* 2010; Lazonick and Tulum 2011).

In my view, this failure to meet expectations might be explained by the particular political economy of the life sciences, which can be summarized as follows:

1 The life sciences sector is underpinned by the monetization and commercialization of knowledge (Pisano 2006), especially in the form of intellectual property (IP). As a result, the introduction of new products and services has become secondary to the capture and exploitation of research by large, multinational pharmaceutical corporations, especially research undertaken by public institutions (e.g. universities) or small life sciences firms that does not represent any risk to shareholder value if it fails.
2 The monetization of knowledge in the form of IP involves a process of *assetization* (or *assetification*) in which knowledge is turned into 'capitalized property' through particular socio-technical configurations (Birch forthcoming). In particular, it involves a financialized organizational configuration represented by the 'dedicated biotech firm' (Mirowski 2012). These firms represent a mechanism for economic rent-seeking via the ownership and out-licensing of IP assets, rather than profit-making via the development of new products and services.
3 The extension and intensification of IP resulting from these changes has actually created a range of problems for the development of new products and services since they restrict access to basic or 'pre-competitive' research. Legal scholars like Michael Heller (2008) argue that this has led to a 'gridlock economy' in which innovation is severely hampered by the rapid expansion of property rights to knowledge.

As a result, the sector has involved the expansion of particular political-economic activities *and* the particular conceptualization of those self-same activities. The combination of these two aspects of the life sciences has led to what I call *rentiership* (Birch *et al.* forthcoming). At base it involves the reconfiguration of property rights and ownership to turn knowledge into assets, and the capitalization of those knowledge assets (Zeller 2008; Birch and Tyfield 2013; Birch forthcoming).

In finishing this book, I want to problematize the very notion that innovation and entrepreneurship underpin the life sciences, despite arguments to the contrary. Rather, the sector is driven by a turn towards *rentiership* in which the creation of value (for firms, customers, investors, regions, etc.) is better theorized as the extraction and enclosure of knowledge. As this is the beginnings of my theoretical engagement with rentiership, I can only outline the bare bones of the concept, but intend to come back to it over the next few years. First, it is premised on the idea that knowledge is not a thing, or even commodity; rather, knowledge is a social process as conceived in social epistemology (Fuller 2013). Consequently, knowledge cannot be analyzed as atomistic, individualistic or similar, since it reflects the 'habits of life' in Veblen's (1908) terms, or 'general intellect' in Marx's.

Second, as a social process, knowledge can only be turned into an asset through the configuration of socio-technical arrangements (e.g. ownership rights, monetization practices). This involves, more often than not, enclosing knowledge to restrict access to it, rather than any productive process. Third, turning knowledge into an asset is better thought of as a process of *assetization* than commodification, driven by property rights and rent-seeking rather production and profit-making (Zeller 2008). Fourth, rent-seeking is not intrinsic to assets, but requires particular political-economic practices and knowledge claims that involve new ways of valuing things (Muniesa 2012; Chiappello 2015). As Chapter 6 illustrates, such valuation is often temporal, involving relay-like relationships amongst investors in which value need not accrue to the initial creative actor. Finally, rentiership is built on specific social forms of organization and governance, especially private firms. Obviously, this notion of rentiership has significant implications for regional development that requires further research and analysis.

References

Amin, A. and Thrift, N. (1992) Neo-Marshallian nodes in global networks, *International Journal of Urban and Regional Research* 16: 571–587.

Andersson, T., Gleadle, P., Haslam, C. and Tsitsianis, N. (2010) Bio-pharma: A financialized business model, *Critical Perspectives in Accounting* 21: 631–641.

Asheim, B. and Gertler, M. (2005) The geography of innovation: Regional innovation systems, in J. Fagerberg, D. Mowery and R. Nelson (eds) *The Oxford Handbook of Innovation*, Oxford: Oxford University Press.

Audretsch, D. and Stephan, P. (1996) Company-scientist locational links: The case of biotechnology, *The American Economic Review* 86: 641–652.

Bair, J. (2005) Global capitalism and commodity chains: Looking back, going forward, *Competition and Change* 9(2): 153–180.

Bathelt, H., Malmberg, A. and Maskell, P. (2004) Clusters and knowledge: Local buzz, global pipelines and the process of knowledge creation, *Progress in Human Geography* 28(1): 31–56.

Birch, K. (2006) The neoliberal underpinnings of the bioeconomy: The ideological discourses and practices of economic competitiveness, *Genomics, Society and Policy* 2(3): 1–15.

Birch, K. (2007) The virtual bioeconomy: The 'failure' of performativity and the implications for bioeconomics, *Distinktion: Scandinavian Journal of Social Theory* 14: 83–99.

Birch, K. (2008) Alliance-driven governance: Applying a global commodity chains approach to the UK biotechnology industry, *Economic Geography* 84(1): 83–103.

Birch, K. (2009) The knowledge–space dynamic in the UK bioeconomy, *Area* 41(3): 273–284.

Birch, K. (2011) 'Weakness' as 'strength' in the Scottish life sciences: Institutional embedding of knowledge-based commodity chains in a less-favoured region, *Growth and Change* 42(1): 71–96.

Birch, K. (forthcoming) Rethinking value in the bio-economy: Corporate governance, assets and the management of value, *Science, Technology and Human Values*.

Birch, K. and Cumbers, A. (2010) Knowledge, space and economic governance: The implications of knowledge-based commodity chains for less-favoured regions, *Environment and Planning A* 42(11): 2581–2601.

Birch, K. and Mykhnenko, V. (2009) Varieties of neoliberalism? Restructuring in large industrially-dependent regions across Western and Eastern Europe, *Journal of Economic Geography* 9(3): 355–380.

Birch, K. and Mykhnenko, V. (2014) Lisbonizing vs. financializing Europe? The Lisbon Strategy and the (un-)making of the European knowledge-based economy, *Environment and Planning C* 32(1): 108–128.

Birch, T. and Tyfield, D. (2013) Theorizing the Bioeconomy: Biovalue, Biocapital, Bioeconomics or… What?, *Science, Technology and Human Values* 38(3): 299–327.

Birch, K., MacKinnon, D. and Cumbers, A. (2010a) Old industrial regions in Europe: A comparative assessment of economic performance, *Regional Studies* 44(1): 35–53.

Birch, K., Levidow, L. and Papaioannou, T. (2010b) *Sustainable Capital?* The neoliberalization of nature and knowledge in the European knowledge-based bio-economy, *Sustainability* 2(9): 2898–2918.

Birch, K., Levidow, L. and Papaioannou, T. (2014) Self-fulfilling prophecies of the European knowledge-based bio-economy: The discursive shaping of institutional and policy frameworks in the bio-pharmaceuticals sector, *Journal of the Knowledge Economy* 5(1): 1–18.

Birch, K., Tyfield, D. and Chiapetta, M. (forthcoming) From neoliberalizing research to researching neoliberalism: STS, *rentiership* and the emergence of commons 2.0, in D. Cahill, M. Konings and M. Cooper (eds) *The SAGE Handbook of Neoliberalism*, London: SAGE.

BIS (2010) *Local Growth: Realising Every Place's Potential*, London: Department for Business, Innovation and Skills.

Callon, M. (ed.) (1998) *Laws of the Markets*, Oxford: Blackwell.

Cerny, P (1999) Reconstructing the political in a globalizing world: States, institutions, actors and governance, in F. Buelens (ed.) *Globalization and the Nation-State*, Cheltenham: Edward Elgar, pp. 89–137.

Chesbrough, H. (2003) *Open Innovation*, Cambridge, MA: Harvard Business School Press.

Chiapello, E. (2015) Financialization of valuation, *Human Studies* 38(1): 13–35.

Coe, N., Lai, K. and Wojcik, D. (2014) Integrating finance into global production networks, *Regional Studies* 48(5): 761–777.

Coenen, L., Moodysson, J. and Asheim, B. (2004) Nodes, networks and proximities: On the knowledge dynamics of the Medicon Valley biotech cluster, *European Planning Studies* 12: 1003–1018.

Cooke, P. (2007) *Growth Cultures*, London: Routledge.

Cooke, P. and Morgan, K. (1998) *The Associational Economy: Firms, Regions, and Innovation*, Oxford: Oxford University Press.

DETR (2000) *Planning for Clusters: A Research Report*, London: Department of the Environment, Transport and the Regions.

DTI (1998) *Our Competitive Future: White Paper*, London: Department of Trade and Industry.

DTI (1999a) *Biotechnology Clusters Report*, London: Department of Trade and Industry.
DTI (1999b) *Genome Valley: The Economic Potential and Strategic Importance of Biotechnology in the UK*, London: Department of Trade and Industry.
DTI (2003) *Innovation Report – Competing in the Global Economy: The Innovation Challenge*, London: Department of Trade and Industry.
Ernst and Young (2015) *Biotechnology Industry Report 2015: Beyond Borders*, Boston MA: EY LLP.
Fagerberg, J. (2005) Innovation: A guide to the literature, in J. Fagerberg, D. Mowery and R. Nelson (eds) *The Oxford Handbook of Innovation*, Oxford: Oxford University Press, pp. 1–26.
Fuller, S. (2013) On commodification and the progress of knowledge in society: A defence, *Spontaneous Generations: A Journal for the History and Philosophy of Science* 7(1): 6–14.
Gereffi, G. (1994) The organization of buyer-driven global commodity chains: How U.S. retailers shape overseas production networks, in G. Gereffi and M. Korzeniewicz (eds) *Commodity Chains and Global Capitalism*, Westport, CT: Greenwood Press, pp. 95–122.
Gereffi, G., Humphrey, J. and Sturgeon, T. (2005) The governance of global value chains, *Review of International Political Economy* 12: 78–104.
Gertler, M. (2003) Tacit knowledge and the economic geography of context, or The undefinable tacitness of being (there), *Journal of Economic Geography* 3: 75–99.
Gertler, M. and Levitte, Y. (2005) Local nodes in global networks: The geography of knowledge flows in biotechnology innovation, *Industry and Innovation* 12: 487–507.
Gertler, M. and Vinodrai, T. (2009) Life sciences and regional innovation: One path or many?, *European Planning Studies* 17(2): 235–261.
Godin, B. (2006) The linear model of innovation: The historical construction of an analytical framework, *Science, Technology and Human Values* 31(6): 639–667.
Heller, M. (2008) *The Gridlock Economy: How Too Much Ownership Wrecks Markets, Stops Innovation, and Costs Lives*, New York: Basic Books.
HM Treasury (2001) *Productivity in the UK: The Regional Dimension*, London: HM Treasury.
HM Treasury (2003) *Lambert Review of Business-University Collaboration*, London: HMSO.
Hopkins, M., Martin, P., Nightingale, P., Kraft, A. and Mahdi, S. (2007) The myth of the biotech revolution: An assessment of technological, clinical and organisational change, *Research Policy* 36(4): 566–589.
Lazonick, W. and Tulum, O. (2011) US biopharmaceutical finance and the sustainability of the biotech business model, *Research Policy* 40(9): 1170–1187.
Leibovitz, J. (2004) 'Embryonic' knowledge-based clusters and cities: The case of biotechnology in Scotland, *Urban Studies* 41: 1133–1155.
Loconto, A. (2010) Sustainability performed: Reconciling global value chain governance and performativity, *Journal of Rural Social Science* 25(3): 193–225.
Lovering, J. (1999) Theory led by policy: The inadequacies of 'The New Regionalism', *International Journal of Urban and Regional Research* 23: 379–395.
Malmberg, A. (2003) Beyond the cluster – Local milieus and global connections, in J. Peck and H. Yeung (eds) *Remaking the Global Economy*, London: SAGE.
Malmberg, A. and Power, D. (2005) (How) do (firms in) clusters create knowledge?, *Industry and Innovation* 12: 409–431.

Mazzucato, M. (2013) *The Entrepreneurial State*, London: Anthem Press.
Mirowski, P. (2011) *ScienceMart*, Cambridge, MA: Harvard University Press.
Mirowski, P. (2012) The modern commercialization of science as a Passel of Ponzi schemes, *Social Epistemology* 26(3–4): 285–310.
Mittra, J. (2016) *The New Health Bioeconomy: R&D Policy and Innovation for the Twenty-first Century*, Basingstoke: Palgrave Macmillan.
Morgan, K. (1997) The learning region: Institutions, innovation and regional renewal, *Regional Studies* 31: 491–503.
Muniesa, F. (2012) A flank movement in the understanding of valuation, *The Sociological Review* 59(s2): 24–38.
OBN (2002) *Growth and Sustainability: The Cluster Report 2002*, Oxford Brookes University: Oxfordshire Bioscience Network.
ODPM (2004) *Our Towns and Cities: The Future*, London: Office of the Deputy Prime Minister.
Owen, G. and Hopkins, M. (2016) *Science, the State and the City*, Oxford: Oxford University Press.
Oxford Trust (2002) *Oxford Networks: Medical and Biosciences Report*, Oxford: The Oxford Trust.
Phelps, N. (2004) Clusters, dispersion and the spaces in between: For an economic geography of the banal, *Urban Studies* 41(5/6): 971–989.
Pike, A., Rodríguez-Pose, A. and Tomaney, J. (2006) *Local and Regional Development*, London: Routledge
Pisano, G. (2006) *Science Business*, Boston: Harvard Business School Press.
Ponte, S. (2009) From fishery to fork: Food safety and sustainability in the 'virtual' Knowledge-Based Bio-Economy (KBBE), *Science as Culture* 18(4): 483–495.
Porter, M. (1990) *The Competitive Advantage of Nations*, London: Macmillan.
Porter, M. (2000) Location, competition, and economic development: Local clusters in a global economy, *Economic Development Quarterly* 14(1): 15–34.
Powell, W., Koput, K., Bowie, J. and Smith-Doerr, L. (2002) The spatial clustering of science and capital: Accounting for biotech firm–venture capital relationships, *Regional Studies* 36(3): 291–305.
Tyfield, D. (2012a) A cultural political economy of research and innovation in an age of crisis, *Minerva* 50: 149–167.
Tyfield, D. (2012b) *The Economics of Science: A Critical Realist Overview* (Volume 1 and 2), London: Routledge.
van Apeldoorn, B. (2000) Transnational class agency and European governance: The case of the European Roundtable of Industrialists, *New Political Economy* 5(2): 157–181.
Veblen, T. (1908) On the nature of capital: Investment, intangible assets, and the pecuniary magnate, *Journal of Economics* 23(1): 104–136.
Walker, R. (2006) The boom and the bombshell: The New Economy bubble and the San Francisco Bay Area, in G. Vertova (ed.) *The Changing Economic Geography of Globalization*, London: Routledge, pp. 121–147.
Zeller, C. (2004) North Atlantic innovative relations of Swiss pharmaceuticals and the proximities with regional biotech arenas, *Economic Geography* 80: 83–111.
Zeller, C. (2008) From the gene to the globe: Extracting rents based on intellectual property monopolies, *Review of International Political Economy* 15(1): 86–115.
Zucker, L., Darby, M. and Brewer, M. (1998) Intellectual human capital and the birth of US biotechnology enterprises, *The American Economic Review* 88: 291–306.

Index

Advisory Board for the Research Councils (ABRC) 43
Advisory Council for Applied Research and Development (ACARD) 6, 43
agglomeration economies 19–20, 24, 41, 49, 82
alliance-driven governance (ADG) 29, 30, 61–3; coordination and trust 72–5; and regional development 130–1, 132; trust 65–6
Amgen 110
Amin, A. 19, 20
Andersson, T. 109, 112, 120, 122
Arora, A. 29, 63
Asheim, B. 25, 49
asset specificity 29, 63, 108
assetization 109, 137, 138
associational theories 21
AstraZeneca 121

bankruptcies 117, 118, 119
Bathelt, H. 8, 21, 54, 73, 74, 82, 83
Bell, D. 1, 2, 17
Berkshire: knowledge and spatial trends 49–51; knowledge–space dynamic 52–4; as regional centre 46–9
bio-economy 7, 93
biopharmaceutical innovation 93–5; and institutional/policy changes 99–101; knowledge-based bio-economy agenda 95–9
biotechnology 5–6
Biotechnology and Biological Sciences Research Council (BBSRC) 44, 50
Birch, K. 2, 4, 8, 9, 10, 21, 25, 27, 28, 29, 40, 54, 61, 62, 63, 67, 71, 72, 86, 90, 91, 106, 108, 109, 112, 113, 114, 115, 117, 119, 121, 127, 129, 130, 131, 132, 136, 137
book outline 9–11
Borrás, S. 91, 92, 95
Borup, M. 88, 90, 98
Bracyzk, H.-J. 8, 41
British Biotech 44
British Technology Group (BTG) 44
Brown, N. 88, 89, 90, 96, 98, 101
Buckinghamshire: knowledge and spatial trends 49–51; knowledge–space dynamic 52–4; as regional centre 46–9
business angel investment 109, 136
business partnerships 79, 80, 89, 113
business strategies: impact of global financial crisis 117–22; outsourcing risk 78–82
buyer-driven governance 26, 28–9

Cambridge Antibody Technology 51
Casper, S. 23, 29, 63
Celltech 44, 51
Chandlerian regime 16
Chesbrough, H. 50, 130
Clinical Laboratory Standards Institute (CLSI) 77
cluster approach 24
codified knowledge 39–40, 51–4, 128; clustering 52
Coe, N. 62, 108, 109, 134–5
Coenen, L. 25, 26, 49
Cold War regime 16–17
commercial knowledge 51–4; clustering 53
commercialization of knowledge 4–5
communities of practice 21
competence trust 65, 76, 77
competitiveness: EU 91–3, 94, 95–8, 101, 133; threats from globalization 60–1; UK 43, 127–8, 129
Cooke, P. 15, 20, 21, 22, 25, 26, 28, 106, 108
creative destruction 18

definitions: bio-economy 7; biotechnology 6; financialization 108; imaginaries 87–8; innovation 4–5, 18; knowledge economy 2–4; life sciences 5–7; regional development 126
Department of Trade and Industry (DTI) 6, 44, 45, 133
drug attrition rates 94–5, 96, 99–100
drug pricing 95, 97, 99, 100, 101

early stage funding 71, 113, 135
East Anglia: knowledge and spatial trends 49–51; knowledge–space dynamic 52–4; as regional centre 46–9; transition to knowledge economy 128
Eastern Scotland: knowledge and spatial trends 49–51; knowledge–space dynamic 52–4; as regional centre 46–9; transition to knowledge economy 128
economic change 1–2
economic development, role of knowledge and innovation 16–18
economic geography 8, 9, 10, 16, 19, 24, 25, 39
economic goals: distinction with societal goals 88, 133; framing with societal goals 60–1, 90–1, 96–7, 98
economic sociology 22, 24, 26–7
economic–political context: EU 60–1; UK 43–5
economies of scale 67, 75, 92
enabling technology 6; role of expectations 88–9
entrepreneurship 18
Enzymatix 44
European Association for Bioindustries (EuropaBio) 101
European Federation of Pharmaceutical Industries and Associations (EFPIA) 94, 96–7, 100
European Patent Office 50, 96, 97
European Technology Platforms (ETPs) 91, 93–4, 97
European Union (EU): *Action Plan for the Life Sciences* 98; agri-food sector 91; *Aho* report 2006 92–3; Cologne Summit 2007 97; definition of life sciences 6, 7; Directorate-General for Research, Science and Education 91–2; Framework Programmes 91, 92, 94; *Growth, Competitiveness, Employment* White Paper 92; *Horizon 2020* 4; implications of research findings 134; Innovation Union programme 4; Innovative Medicines Initiative (IMI) 93–101; institutional and policy changes 99–101, 133; knowledge-based bio-economy strategy 8–9, 86–7, 91–3; knowledge-based bio-economy strategy as imaginary 93–5; 'lagging behind' diagnosis 95–7, 133; *Life Sciences Strategy* 93; *Lisbon Agenda* 4, 60–1, 86–7, 92, 93, 97, 107–8; policy discourse 60–1, 107–8; regional development and imaginaries 132–4; research and innovation policy 91–3, 133; social model 60–1; societal responses to research bottlenecks 97–9; Strategic Research Agendas 91, 94, 95, 96–7; *Strategy for Europe on Life Sciences and Technology* 6
evolutionary economics 23, 24, 25, 41
exit opportunities, impact of global financial crisis 120–1
extra–local linkages 25–6, 61–3, 131; coordination and trust 72–5; governance of collaboration 75–8; governance of risk outsourcing 78–82; Scotland 68–71

Fagerberg, J. 18, 20
Feldman, M. 19, 49
fetishization of innovation 101, 107, 120, 127, 133, 134, 136
finance: capital availability 118–19; changes in business strategies 117–22; early research 106–7; early stage funding 71, 113; EU 94–5; global IPOs 115; and global life sciences 110–12; impact of global financial crisis 115–22; later stage funding 113–14; market capitalization, revenues and profit 111; methodological note 109–10; new forms of 63; private investment 115; public funding 29, 44, 50, 67, 71, 95–7, 112–13, 119, 135; and regional development 134–6; 'relay-like' process of 109, 120, 122; returns on investments 110–11; revenues and profit/loss by firm size 112; role of technological expectations 88–9; Scotland 71, 113, 119; seed and pre-seed investment 112–13; specialist knowledge 119; specific types of relationship with life sciences 108–9; strategic investing 121–2; UK firms on public markets 116; and uneven development 136; valuation practices 119, 120–1
financial ecosystem: impact of global financial crisis 119; and innovation 135

financialization 107–9, 135, 136–7; definitions 108
firm-centred approaches 18, 39, 40–1
Fuller, S. 4, 138
functional theories 19–20

Genentech 16, 110
geographical imaginaries 87–8, 133–4
geography of innovation 18–21
Gereffi, G. 8, 10, 26, 27, 29, 62
Gertler, M. 83, 128, 129, 134
Gillespie, I. 98, 99
GlaxoSmithKline 113, 121
global commodity chain (GCC) approach 8, 26–7, 28, 61, 62–3; analytical benefits 27; governance of innovation outsourcing 78–82; governance of innovation relations 75–8; positioning along 67–8, 131; and trust 72–5
global financial crisis (GFC) 108, 110, 113, 135; impact on capital availability 118–19; impact on exit opportunities 120–1; impact on financial ecosystem 119; impact on innovation promises 117–18; impact on investment 114–17; impact on R&D strategies 121–2
global production networks (GPNs) 27–8
global value chain (GVC) approach 8, 10, 27, 62, 135
globalized privatisation regime 17
Godin, B. 2, 4–5, 17, 18
goodwill trust 65, 76, 77
Gottweis, H. 43, 44, 92
governance: alliance-driven 61–3, 130–1; and coordination and trust 72–5; of innovation relations 78–82; of innovation relations and collaboration 75–8; knowledge-based commodity chains in less-favoured regions 68–71; methodological note 66; models 26, 28–9; positioning Scotland in commodity chains 67–8; and regional development 130–2; and trust in knowledge-based commodity chains 63–6
Gray, M. 25–6, 76

healthcare remuneration policies 136
Hopkins, M. 43, 110, 112, 113, 114, 119

imaginaries: and biotech clusters 132–3; definition 87–8; descriptive accounts 87, 93, 95–7, 133; and EU research and innovation policy 91–3; implications of research findings 134; KBBE agenda as 93–5; methodological note 91; normative accounts 87, 93, 97–9, 133; performative accounts 87, 89, 90–1, 93, 99–101, 132, 133; and regional development 132–4; as self-fulfilling prophecies 89–91; and technological expectations 88–9
incentives, coordination of 29, 63, 72, 77–8, 80, 81–2, 88
initial public offerings (IPOs) 113–15, 135; impact of financial crisis 120–1
innovation gap, EU 91–3
innovation promises, impact of global financial crisis 117–18
innovation studies 4–5, 8, 16, 17–18, 20, 22–3, 24, 39
institutional frameworks 28; as cause of R&D bottlenecks 95–7; changes in 92–3, 97–8, 99–101; and knowledge-based bio-economy agenda 93–7; role of imaginaries 89–91; UK 50–1; weaknesses as potential strengths 132
institutional investment 109, 120
intellectual property (IP): monetization of 109, 121, 136–7; new forms of 29, 63; policies 96, 97; protection and definition of 65, 80–2
inter-disciplinary working 72–3
inter-organizational relationships 22, 23, 42, 50, 69–70, 72, 82–3
International Standards Organization (ISO) 65, 76–7
iterative learning 42–3

Jasanoff, S. 86, 87
Jessop, B. 1, 2, 86, 87, 88

Kettler, H. 29, 63
knowledge exchange, governance of 75–8
knowledge spillovers 23, 24, 25, 29, 41, 45, 49, 50–1, 54, 128, 129
knowledge-based bio-economy (KBBE) 86–7; context 91–3; European Commission agenda 94; European Commission agenda as imaginary 93–5; and institution/policy changes 99–101; promotion of societal responses 97–9
knowledge-based commodity chains 16, 28–9, 130; governance 61–6; governance in Scotland 72–82; in less-favoured regions 68–71; positioning Scotland 67–8; 'stickiness' 63
knowledge-based economy (KBE) 15, 17, 60–1; definitions 2–4; UK 128

knowledge-based theories 22–3
knowledge–space dynamic: knowledge and spatial trends in UK 49–51; knowledge processes 40–1; methodological note 45–6; political–economic context in UK 43–5; regional centres in UK 46–9; and regional development 127–30; spatial processes 41–2; synthesizing debates 42–3; in UK life sciences 51–4

Lagendijk, A. 19, 20, 21
large firms: 'anchoring' role 49–50; globalization of R&D 92; in governance of risk outsourcing 78–82; in governance of collaboration 75–8; revenues 110–11
Lazonick, W. 108, 121
lead market initiatives 101
less-favoured regions (LFRs): and EU policy 60–1; implications of trust in alliance-driven governance 65–6; in knowledge-based commodity chains 67, 68; regional development 130, 131, 132
Levidow, L. 8, 87, 88, 91
Leydesdorff, L. 15, 25, 41
licensing 79, 80, 109, 137
life sciences sector 15–16; definition 5–7; peculiar characteristics 136–8
linear innovation model 3–4, 10, 17, 27, 130
listing 44; and delisting 116–17, 119
local-boundedness 62, 69–70
London Stock Exchange (LSE) 44
London: knowledge and spatial trends 49–51; knowledge–space dynamic 52–4; as regional centre 46–9; transition to knowledge economy 128

MacKinnon, D. 19, 65
mainstream economics 5, 16, 17, 60, 65
Malmberg, A. 26, 51, 54, 62, 128
manufacturing sector 1–2
market failure 95, 97, 100, 101
markets, lack of 131, 132
Markusen, A. 5, 63
Mazzucato, M. 113, 135
Medical Research Council (MRC) 44, 50
Merges, R. 29, 63
methodological framework 7–9; finance 109–10; governance 66; imaginaries 91; knowledge–space dynamic 45–6
Michael, M. 88, 89, 96, 98, 101
Mirowski, P. 3, 16, 109, 137
modern biotechnology 6
Moodysson, J. 21, 79
Morgan, K. 20, 21
Moulaert, F. 19, 20
multi-disciplinary working 72–3
multi-scalar processes 10, 16, 21, 25, 26, 28, 29, 45, 46–61, 70, 71, 77–8, 129
Mykhnenko, V. 2, 4, 86, 107–8

National Enterprise Board (NEB) 44
National Research Development Corporation (NRDC) 44
Natural and Environmental Research Council (NERC) 50
new economic geography (NEG) 23, 24, 25, 41
new industrial spaces (NIS) 19–20
niche expertise 79

objective trust 65, 76–8
open innovation 22, 130–1
opportunity costs 63, 79, 83, 99–100, 135
Organization for Economic Co-operation and Development (OECD): definition of life sciences 7; promotion of bio-economy policy agenda 93
Ossenbrugge, J. 29, 63
Owen, G. 43, 44
Oxfordshire: knowledge and spatial trends 49–51; knowledge–space dynamic 52–4; as regional centre 46–9; transition to knowledge economy 128

Parker, E. 25–6, 76
patents 50–1, 80–1, 86, 95, 97, 101, 102
'patient' capital 112
Phelps, N. 62, 129
Pike, A. 108, 126
Pisano, G. 79, 106, 109, 110, 118, 121, 137
pharmaceutical productivity crisis 136–7
Polanyi, M. 3, 39–40
policy frameworks: changes in 97–8, 99–101, 133; concerns 136–7; EU 91–3;and knowledge-based bio-economy agenda 93–7; as cause of R&D bottlenecks 95–7; role of imaginaries 89–91
political context: knowledge and innovation 16–18; UK 43–5, 129, 133–4
Porter, M. 1, 8, 20, 21, 24, 39, 41, 129, 132–3
Powell, W. 22, 40, 108, 136
Power, D. 54, 62, 128
pre-competitive research 94–6, 98–9, 100, 102, 137; as barrier to innovation 99–100; subsidies 99

producer-driven governance 26, 28–9
product approval 95, 97, 98, 100
public funding 29, 112–13, 135; impact of financial crisis 120; Scotland 67, 71; systemic failures of EU companies 95–7; UK 44, 50
public good, innovation as 3, 4
public–private partnerships 44, 94–5, 96, 98, 99–100, 101

quality assurance 77

regional development: and finance 134–6; and governance 130–2; and imaginaries 132–4; implications of research findings 129–30, 132, 134; and knowledge–space dynamic 127–30; problem of innovation 136–8
regional innovation systems (RIS) 20, 24, 25, 41–2
regional knowledge capabilities 25, 26, 51, 62
regional studies 8, 9, 10, 16, 19, 23, 24, 25, 41, 129
regulatory compliance, and trust 65, 76–8
Reich, R. 1, 60
relational theories 19, 20–1
rentiership 137–8
research and development (R&D): EU bottlenecks 93–5; EU technology gap 91–3; finance 94–5, 97; impact of global financial crisis 121–2; importance of technological expectations 89; and institutional/policy changes 99–101; outsourcing 121; societal responses to bottlenecks 97–9; systemic failures 95–7
research and innovation policy 2–3
risk outsourcing, governance of 74–5, 78–82, 135
Rosiello, A. 61, 67, 71

Sanz-Menéndez, L. 91, 92
Schumpeter, J. 4, 17–18, 23, 41, 42
Science and Engineering Research Council (SERC) 44
science and technology (STS) studies 87, 88–90, 91, 93, 98, 99, 101, 102, 106
science organization regimes 16–17
science-based relationships 70
Scotland: alliance-driven governance 130–1; commodity chain relationships 69; finance 71, 113, 119, 136; emergence of life sciences 67; geographies of commodity chain relationships 70; governance of innovation outsourcing 78–82; governance of innovation relations and collaboration 75–8; governance, coordination and trust 72–5; implications of research findings 132; inter-organizational relationships 69; knowledge-based commodity chains in a less-favoured region 68–71, 131; local-boundedness 69–70; methodological note 66; positioning in commodity chains 67–8, 131; public sector support 67; relationships with public sector organizations 71; uneven development 66, 67; *see also* Eastern Scotland
Scott, A.19, 20
Scottish Enterprise 67, 113
seed/pre-seed finance 112–13
Sekia, F. 19, 20
self-fulfilling prophecies 89–91
Senker, J. 23–4, 29, 63
Sharp, M. 23, 44
Silicon Valley 19–20, 130
small firms: exit opportunities 120; in governance of collaboration 75–8; in governance of risk outsourcing 78–82, 121; trading data 117
social capital 25, 42
societal goals: distinction with economic goals 88, 133; framing with economic goals 60–1, 90–1, 96–7, 98
societal responses, research bottlenecks 97–9
Sokol, M. 2, 109
specialized knowledge: capital markets 119; developing and protecting 74–5, 78–82; diversifying to avoid risks of failure 79; ‘upscaling’ across organizational relationships 75–6
spin-offs 16, 67
Staffas, L. 5, 7
‘star scientists’ 24, 47
Storper, M. 20, 65
supplier relationships 70, 76–7
system-centred approaches 18–19, 23, 39, 40, 41
systemic failure 95, 97, 99, 100

tacit knowledge 24, 25, 39–40, 42, 51–4, 128; clustering 53; problem of 3
takeovers 79–80
technological determinism 89–90. 98–9
technological expectations 88–9, 98; performative and realist conceptions 90–1

territorial innovation models (TIMs) 19, 21
Thrift, N. 19, 20, 108
trade sales 113, 114, 115, 120–1, 122, 135
trans-disciplinary working 72–3
Triple Helix model 41
trust 42, 132; and coordination 72–5; definition of 63, 65; in knowledge-based commodity chains 63–6; in regulatory compliance 76–8; in risk outsourcing 79–80, 81–2; types of 65

UK: Bioscience Innovation and Growth Team (BIGT) 6; *Biotech Clusters and Genome Valley* reports 133; Business Funding Zones 129; competitiveness debate 43, 127–8, 129; concentration of larger firms 49–50; Conservative government 43–4, 127, 133–4; definition of life sciences 6; Enterprise Management Incentive Scheme (EMIS) 45; finance 109, 115–17, 122, 136; firm changes on public markets 116; formal alliances and multi-scalar dimensions 45, 46–61; geographical imaginaries 133–4; global alliances 51; implications of research findings 129–30; Innovative Clusters Fund 129; knowledge and spatial trends 49–51; knowledge assets 45, 46–51; knowledge economy 128; knowledge–space characteristics 48; knowledge–space dynamic 51–4; Labour government 6, 43, 45, 127–8; *Local Growth* policy 2010 133–4; localized knowledge, learning and innovation 128–9; national alliances 51; one-size-fits-all policy 129, 134; *Our Plan for growth* 3; *Planning Policy Guidance* (PPG) 45; political–economic context 43–5, 133–4; public science base 45, 46–51, 55; regional centres 46–9; Regional Development Agencies (RDAs) 45, 127, 129, 134; regional development implications of uneven life sciences spread 55; Regional Innovation Fund 129; regional institutional ecosystem 50–1; Regional Venture Capital (RVC) Funds 45; size and extent of biotech firms 45, 46–51; *Spinks Report* 6, 43, 44, 136; *Strategy for UK Life Sciences* 6; White Paper 1981 43–4
uneven development 2, 19, 39, 40, 46, 51, 54, 55, 66, 67, 107, 128, 129–30, 131; and finance 136
US Patent Office 50
US: biopharmaceutical sector 97; as dominant market 131; as exemplar 134, 135; financial market 107; shift to knowledge economy 17

valuation practices 119, 120–1
van Lente, H. 88, 102
varieties of capitalism (VOC) approach 23–4
venture capital (VC) 16, 19, 29, 45, 63, 66, 71, 106, 108, 110, 112, 113, 122, 130, 134, 135, 136; impact of global financial crisis 119
Vinodrai 129, 134

willingness to pay 99, 100

Zeller, C. 29, 51, 63, 138

For Product Safety Concerns and Information please contact our EU
representative GPSR@taylorandfrancis.com
Taylor & Francis Verlag GmbH, Kaufingerstraße 24, 80331 München, Germany

www.ingramcontent.com/pod-product-compliance
Ingram Content Group UK Ltd.
Pitfield, Milton Keynes, MK11 3LW, UK
UKHW020948180425
457613UK00019B/584

* 9 7 8 0 3 6 7 8 7 0 8 9 8 *